The Smart Enough City

** Ideas Series**

Edited by David Weinberger

The Ideas Series explores the latest ideas about how technology is affecting culture, business, science, and everyday life. Written for general readers by leading technology thinkers and makers, books in this series advance provocative hypotheses about the meaning of new technologies for contemporary society.

The Ideas Series is published with the generous support of the MIT Libraries.

Hacking Life: Systematized Living and Its Discontents, Joseph M. Reagle, Jr.

The Smart Enough City: Putting Technology in Its Place to Reclaim Our Urban Future, Ben Green

The Smart Enough City

Putting Technology in Its Place to Reclaim Our Urban Future

Ben Green

The MIT Press
Cambridge, Massachusetts
London, England

First MIT Press paperback edition, 2020
© 2019 Massachusetts Institute of Technology

This book was set in Stone Serif and Stone Sans by Jen Jackowitz. Printed and bound in the United States of America.

Library of Congress Cataloging-in-Publication Data

Names: Green, Ben (City planner), author.
Title: The smart enough city : putting technology in Its place to reclaim our
    urban future / Ben Green.
Description: Cambridge, MA : The MIT Press, [2019] | Series: Strong ideas |
    Includes bibliographical references and index.
Identifiers: LCCN 2018030979 | ISBN 9780262039673 (hardcover : alk. paper)—
    9780262538961 (paperback)
Subjects: LCSH: Smart cities.
Classification: LCC TD159.4 .G74 2019 | DDC 307.76--dc23 LC record available at
    https://lccn.loc.gov/2018030979

10 9 8 7 6 5

For Pop and Mom

Contents

Foreword

The age of the "Smart City" is upon us! It's just that, we don't really know what that means. Or, at least, not yet.

—The Boston Smart City Playbook (2016)

As chief information officer of the City of Boston from 2014 to 2018, I sat through many pitches from companies selling "smart cities" technology. A memorable one came from two Fortune 500 companies who had partnered to offer a connected device that would sit on top of our streetlights and provide cameras, sensors, and computing capabilities at tens of thousands of locations across Boston.

Like most smart cities products, it was pitched as a "platform" that would capture many kinds of data and, with the right analytical models, enable us to improve everything from traffic flow to public safety to the efficiency of city services.

A colleague asked the assembled vendors whether any of these benefits had actually been realized in practice, to which one of the company division heads enthusiastically replied, "That's the exciting part about it: we give you the platform and the data and you get to figure out all the ways you can get value from it." The kicker came when we asked about price, and learned that annual service costs alone were almost as much as the city spends on snow removal and trash collection combined. Not a proposal that I was going to rush down to the mayor's office.

The experience highlighted a common gap in the world of the smart city. Companies see possibilities (and dollar signs), while municipal employees see hard financial trade-offs and a complicated path to translate technology into real public value. Moreover, it illustrates a fundamental difference in how people see the challenges facing cities. To technologists, cities are a collection of straightforward optimization problems for which more data and computing power can only be helpful—who could argue with making traffic flow better and delivering services more efficiently?

But to those on the front lines, words like "better" and "more efficient" are the tip of an iceberg, below which sit the competing interests and conflicting values of the city and the people who live in it. Even a simple concept like improving traffic flow quickly breaks down into thorny questions of priority and perspective: *Should we automatically give a green light to a bus approaching an intersection, even if doing so slows down other drivers? Is it fair to retail businesses to take away street parking for an Uber pickup zone? Should we use predictive signal timing to speed up traffic if that might make roads less safe for people who walk and bike?* These are not technical questions, and no amount of sensor data can provide the right answer.

When I joined Boston City Hall after ten years running a technology company that I founded, I learned quickly that sentences which began with *"Couldn't we just . . ."* often ended with my foot in my mouth. People who had worked with a problem for years often schooled me in the complex political and structural challenges at the heart of issues that seemed on the surface to be fixable with a small splash of tech wizardry. The tendency to overlook deeper questions of values and trade-offs—in favor of a reductive, solutionist approach—is one of the blind spots of many technologists.

The Smart Enough City dives deep into the opportunity and challenge of applying technological solutions to the very human-centric world of urban governance. Ben Green articulates the incredible promise that new technologies offer cities, while embracing the complications and complexity that they bring in implementation. By rejecting the tech-centric thinking that undergirds many smart cities concepts, he shows us how to avoid the trap of seemingly simple solutions to wicked problems.

Ben also posits a new role for the practitioners of technology in government, one that he modeled in his time working with the City of Boston. In his year as a data scientist with my team at the Department of Innovation and Technology, Ben was a thought partner on the role of innovation and the thorny complexity of its impacts on city residents. On topics spanning public Wi-Fi to sidewalk repairs, he helped the city carefully weave the values and priorities of our community into the application of exciting new technology. And in his hands-on work with city departments, he pushed beyond the false promise of superficial optimization and zero-trade-off improvements.

As someone with deep data science skills, Ben was asked to support Boston Emergency Medical Services (EMS) in addressing a troubling increase in ambulance response times. His approach was one of inquiry. He analyzed data about usage trends, call types, and ambulance geography. As importantly, he built relationships with EMS leadership and paramedics on the ground. An ambulance ride-along put a human face on the patients being served and offered context on the factors that influence response time. Modeling data alongside the people whose work it represented led to some impactful insights.

He was able to show how and where ambulance capacity had failed to keep up with demand, and his thoughtful analysis uncovered opportunities to improve the EMS service model. A significant amount of ambulance time was used by nonmedical emergency calls for people struggling with homelessness, drug addiction, or both. EMS responders were acting as de facto social service providers—a critically important function, but not one best performed by paramedics on a fully equipped ambulance.

Ben's people-focused, collaborative approach led to the creation of an EMS Community Assistance Team that works with substance abuse and outreach workers to connect people in crisis to social services. Responders with special training and resources were able to offer better care and support, while freeing up ambulance units for calls that required their unique skills and equipment.

The Smart Enough City busts the myth of the swashbuckling urban innovator reshaping the city with the simple magic of disruptive technology. It offers both a warning and a road map to those who make it their business to make cities smarter. Technologists can have big positive impacts on cities only by marrying deep technical skills with careful program design and an empathetic embrace of the complexity and contradictions of city life. For those of us optimistic about the potential to develop new solutions to long-standing urban challenges, this book brings an important new voice and a thoughtful path forward.

Jascha Franklin-Hodge
Former Chief Information Officer, City of Boston
Co-founder, Blue State Digital

Acknowledgments

Many people are responsible for making this book what it is.

This book would never have been possible without the mentorship provided by Susan Crawford. The first person to truly recognize my mishmash of interests—spanning machine learning to urban policy—as an asset, Susan provided me with far more agency and responsibility than I deserved. Her personal and financial support was a wonderful gift.

I am indebted to many other mentors who have opened my eyes to a variety of issues at stake in technology and urban policy, and helped develop my critical eye: Yochai Benkler, Julia Freeland, Rayid Ghani, Michelle Mangan, Radhika Nagpal, Andrew Papachristos, Todd Reisz, Jim Travers, and Mitch Weiss.

The Berkman Klein Center for Internet & Society at Harvard has been a wonderful hub of friendship and intellectual inspiration. It is an honor to be a part of this extraordinary community, and I am grateful in particular for those who helped me develop the projects and ideas that led to this book: David Cruz, Gabe Cunningham, Ariel Ekblaw, Paul Kominers, Andrew Linzer, Jon Murley, Maria Smith, Dave Talbot, and Waide Warner.

My experience working for Boston's Department of Innovation and Technology was an incredibly formative one, and in large part provided the backdrop for this book. I owe a great thanks to Jeff Liebman for granting me the unique opportunity to spend a year there. Patricia Boyle-McKenna, Jascha Franklin-Hodge, and Andrew Therriault were great leaders who created an atmosphere of service and innovation. Many others were wonderful mentors, colleagues, and friends: Alex Chen, Stefanie Costa-Leabo, Elijah de la Campa, Chris Dwelley, Joseph Finn, Peter Ganong, Jim Hooley, Nigel Jacob, Ramandeep Josen, Kayla Larkin, Howard Lim, Sam Lovison, Kim Lucas, Laura Melle, Chris Osgood, Kayla Patel, Jean-Louis Rochet, Luis

Sano-Espinosa, Anne Schwieger, Steve Stephanou, Renee Walsh, Steve Walter, and all of Intern Island.

Stanford's Center for Internet and Society was a wonderful place to spend a summer writing retreat. Thank you to everyone there, and especially Al Gidari, for graciously welcoming me.

My remarkable series editor, David Weinberger, helped transform my rough sketches of an idea into a fully developed book outline. He pushed me to hone my argument and convey it effectively. This project would never have gotten off the ground without his thoughtful aid.

The entire team at the MIT Press has been a treat to work with. I am deeply grateful for Gita Manaktala's willingness to take a chance on a young and unproven writer, and her unflagging support. Through their constructive feedback, three anonymous reviewers gave me new perspectives on the book's contributions and limitations. Alice Falk provided expert and meticulous copyediting.

The book was vastly improved by the thoughtful commentary and revisions provided by two fantastic editors. Chloe Fox helped me weave meaningful narratives and taught me a great deal about how to structure this project. Ciarán Finlayson provided incisive commentary that kept my arguments sharp. With their aid, I was able to turn my early manuscripts into something resembling a Real Book.

I have benefited from several friends who volunteered their time to review working drafts of the book: Varoon Bashyakarla, Evan Green, Ben Lempert, Robert Manduca, and Drew Ohringer. Zach Wehrwein introduced me to many of the books and ideas that provided an intellectual foundation for this project.

My parents, Jenny Altshuler and Barry Green, have always been my greatest teachers and supporters. Whether on vacation or driving to school, they never passed up an opportunity to demonstrate the value of education. My interdisciplinary approach stems from their example of lifelong learning. Through trips and long, meandering walks, they inculcated me with a love of cities from a young age.

Finally, I am overwhelmed with gratitude for my incredible partner, Salomé Viljoen. Her fingerprints are everywhere in the book, from its intellectual framing to the structure of individual sentences to the fact of its completion. She remained remarkably patient and supportive through my many long nights and weekends of work, not to mention the countless drafts that I asked her to read. This book is a testament to her love.

1   The Smart City: A New Era on the Horizon

In 2016, an article appeared in the *Boston Globe* with a headline that city drivers everywhere have dreamed of saying: "Bye-bye traffic lights."[1] No, Boston had not suddenly removed every traffic light in the city. But such a change appeared to be around the corner: researchers at the Massachusetts Institute of Technology (MIT) had devised new "intelligent intersections"[2] that would enable oncoming streams of self-driving cars to merge seamlessly and travel through intersections without stopping.[3] Once this new technology was deployed, sitting in traffic would be a relic of the past. Simulated demonstrations of these futuristic streets seemed to augur the dawn of a new era, in which advanced technology would alleviate issues that had long plagued cities.

But there was something missing from the mathematical models and simulations that MIT had devised: people. Their city streets showed no sign of life beyond the flow of cars. What makes this omission particularly notable is that the intersection at the heart of the MIT models is among the busiest pedestrian and transit thoroughfares in downtown Boston and one of the most walkable locations in the entire United States.[4] Nobody likes traffic, but if eliminating it requires removing people from streets, what kinds of cities are we poised to create?

These MIT researchers were neither the first nor last to imagine the remarkable benefits that technological advances would bring to cities. Each proposal sounds magnificent, but if you simply scratch the surface of these futuristic models and utopian promises, a more ominous story emerges.

Consider "predictive policing"—machine learning algorithms that analyze historical crime patterns to predict when and where the next crimes will occur. With this information, many believe, police officers can efficiently prevent crime and make communities safer. These algorithms seem to provide an objective and scientific way to maximize limited police resources. Police departments across the United States have adopted the software over the past decade, with one police chief hailing it for helping "us get smarter on our fight against crime."[5]

But these algorithms have a dark side: the information that guides their predictions is imbued with racial bias. Instead of representing an objective reality of where every crime has occurred, the data indicates where police have observed and prosecuted crimes—information that reflects the disparate ways that police treat different communities. By relying on this data, predictive policing software overestimates crime in minority neighborhoods and underestimates crime in white neighborhoods. Acting on these predictions exacerbates the existing biases in policing. Nobody wants crime, but if preventing it means perpetuating discriminatory practices, what kinds of cities are we poised to create?

Consider one more story. In 2016, New York City replaced thousands of pay phones with digital kiosks to create the world's largest and fastest free public Wi-Fi network.[6] These kiosks, branded under the name LinkNYC, also offer free domestic phone calls, USB charging ports, and interactive maps. As an added bonus, the kiosks do not cost the city a dime. LinkNYC's deployment highlighted the need for every city to democratize access to high-speed internet.

Once again, however, this new technology came with caveats. The LinkNYC kiosks are not a public service run by New York City. Instead, they are owned and operated by Sidewalk Labs—a subsidiary of Alphabet, the parent company of Google. How the kiosks are actually funded should thus be no surprise: Sidewalk Labs gathers data about everyone who uses the services, enabling it to generate targeted advertising. Connecting to the public Wi-Fi network therefore comes at the cost of providing data about your location and behavior to private companies. Everybody desires better public services, but if deploying them entails setting up corporate surveillance nodes throughout urban centers, what kinds of cities are we poised to create?

Each of these stories points toward a new type of city that is on the horizon, made possible by new technology: the "smart city." This book is about why, far too often, applications of technology in cities produce adverse consequences—and what we must do to ensure that technology helps create a more just and equitable urban future.

* * *

Thanks to the development of new technologies that make previously unimaginable capabilities routine, cities appear to be on the brink of a revolutionary breakthrough. We are promised that the benefits of these technologies—and the "smart cities" they help create—will be tremendous. Everyday objects will be embedded with sensors that can monitor the world around them. Machine learning algorithms will use this data to predict events before they occur, and to optimize municipal services for efficiency and convenience. Through apps, algorithms, and artificial intelligence, new technology will relieve congestion, restore democracy, prevent crime, and create free public services. The smart city will be the city of our dreams.

From major technology companies to the Obama White House to the National League of Cities,[7] the smart city has garnered widespread support and emerged as the consensus vision for the future of municipal governance. A 2016 survey of fifty-four U.S. cities found that they had collectively implemented or planned almost 800 smart city projects.[8]

Here's how the CEO and vice president of the technology company Cisco describes where we are heading: "By definition, Smart Cities are those that integrate information communications technology across three or more functional areas. More simply put, a Smart City is one that combines traditional infrastructure (roads, buildings, and so on) with technology to enrich the lives of its citizens."[9]

This general description—applying data and technology to traditional objects or processes to enhance efficiency and convenience—has come to define what it means to make something "smart," in cities and beyond. It is in this sense, as a term of art, that I will employ the word throughout the book.

Yet the promises of smart cities are illusory. Their deception stems from their very definition, which overemphasizes the power and importance of technology. Notice how Cisco grounds urban progress solely in the application of technology. This same focus is what produced the dangers of "intelligent intersections," predictive policing, and LinkNYC (examples that we will return to later in the book). As we will see, the problem with smart cities is not merely that technology is incapable of generating the promised benefits but also that attempts to deploy technology in pursuit of a smart city often distort and exacerbate the problems that are supposedly being solved.

Although presented as utopian, the smart city in fact represents a drastic and myopic reconceptualization of cities into technology problems. Reconstructing the foundations of urban life and municipal governance in accordance with this perspective will lead to cities that are superficially smart but under the surface are rife with injustice and inequity. The smart city threatens to be a place where self-driving cars have the run of downtowns and force out pedestrians, where civic engagement is limited to requesting services through an app, where police use algorithms to justify and perpetuate racist practices, and where governments and companies surveil public space to control behavior.

Technology can be a valuable tool to promote social change, but a technology-driven approach to social progress is doomed from the outset to provide limited benefits or beget unintended negative consequences. As the philosopher John Dewey wrote, "The way in which [a] problem is conceived decides what specific suggestions are entertained and which are dismissed."[10] The sociologist Bruno Latour adds, "Change the instruments, and you will change the entire social theory that goes with them."[11] Dewey's and Latour's logic highlights where dreams of the smart city go astray: when we conceive of every issue as a technology problem, we entertain technical solutions and dismiss other remedies, ultimately arriving at narrow conceptions of what a city can and should be.

I call this perspective "technology goggles" (or simply "tech goggles"). At their core, tech goggles are grounded in two beliefs: first, that technology provides neutral and optimal solutions to social problems, and second, that technology is the primary mechanism of social change. Obscuring all barriers stemming from social and political dynamics, they cause whoever wears them to perceive every ailment of urban life as a technology problem and to selectively diagnose only issues that technology can solve. People wearing tech goggles thus perceive urban challenges related to topics such as civic engagement, urban design, and criminal justice as being the result of inefficiencies that technology can ameliorate, and they believe that the solution to every issue is to become "smart"—internet-connected, data-driven, and informed by algorithms—all in the name of efficiency and convenience. Seeing technology as the primary variable that can or should be altered, technophiles overlook other goals, such as reforming policy and shifting political power.

The fundamental problem with tech goggles is that neat solutions to complex social issues are rarely, if ever, possible. The urban designers Horst Rittel and Melvin Webber describe urban social issues as "wicked problems," so complex and devoid of value-free, true-false answers that "it makes no sense to talk about 'optimal solutions.'"[12] Suggesting that technology can solve these types of problems—an attitude that the technology critic Evgeny Morozov decries as "solutionism"[13]—is misguided at best and duplicitous at worst.

Tech goggles do more than merely generate well-intended but ineffectual gizmos, however—they engender a dangerous ideology that has the potential to reshape society. Through a process that I call the "tech goggles cycle," tech goggles warp behaviors, priorities, and policies according to the logic of technology. The cycle operates in three stages. First, *tech goggles* create the perception that every issue can and should be solved with technology. This perspective leads people, companies, and governments to develop and adopt new *technology* intended to make society more efficient and "smart." As municipalities and urban residents adopt this technology, their behaviors, beliefs, and policies are shaped by the misguided assumptions and priorities embodied in these artifacts—*reinforcing* the perspective of tech goggles and bolstering the technologies shaped in their image. Through this process, alternative goals and visions that are not grounded in technology become harder both to recognize and to act on. The perspective of tech goggles becomes more deeply entrenched in our collective imagination.

Embedded in these technologies, and the social changes they beget, is politics. For technologies are not mere neutral tools. As the political theorist Langdon Winner explains in *The Whale and the Reactor*, technologies "embody specific forms of power and authority." Winner adds:

> technological innovations are similar to legislative acts or political foundings that establish a framework for public order that will endure over many generations. For that reason, the same careful attention one would give to the rules, roles,

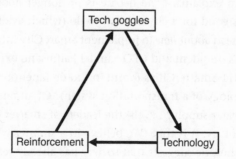

Figure 1.1
The tech goggles cycle.

and relationships of politics must also be given to such things as the building of highways, the creation of television networks, and the tailoring of seemingly insignificant features on new machines. The issues that divide or unite people in society are settled not only in the institutions and practices of politics proper, but also, and less obviously, in tangible arrangements of steel and concrete, wires and semiconductors, nuts and bolts.[14]

Cities cannot escape the need to grapple with values and politics by adopting newer and more efficient technologies. The ways in which we develop and deploy smart city technologies will have vast political consequences: who gains political influence, how neighborhoods are policed, who loses their privacy. Yet tech goggles cause their devotees to perceive complex, normative, and eternally agonistic political decisions as reducible to objective, technical solutions. By conceptualizing urban issues as technology problems, smart city ideologues lose sight of these issues' normative and political elements. In turn, they evaluate solutions along technical criteria (such as efficiency) and overlook the broader consequences.

As Adam Greenfield, who presented one of the earliest and most trenchant critiques of smart cities in his 2013 book *Against the Smart City*, explains, such thinking "is effectively an argument [that] there is one and only one universal and transcendently correct solution to each identified individual or collective human need; that this solution can be arrived at algorithmically, via the operations of a technical system furnished with the proper inputs; and that this solution is something which can be encoded in public policy, again without distortion."[15]

This logic makes the smart city appear value-neutral and universally beneficial—as if it were the only reasonable way forward. Cisco's Urban Innovation team explains, "The debate is no longer about *why* a Smart City initiative is good for a city or *what* to do (which available options to choose), but instead about *how* to implement Smart City infrastructures and services."[16] IBM's president and CEO Samuel Palmisano expressed a similar position at a 2011 SmarterCities forum in Rio de Janeiro: "Think about it. What is the ideology of a transportation system? Of an energy grid? Of an urban food or water supply? . . . [If] the leaders of smarter city systems . . . do share an ideology, it is this: 'We believe in a smarter way to get things done.'"[17] Such rhetoric suggests that society has already reached a consensus about what type of cities to pursue, or perhaps that such a consensus can simply be assumed owing to the splendor of smart city possibilities. To technologists, the benefits of enhanced efficiency are so obvious that the smart city transcends social and political debate—nay, renders it obsolete.

Of course, it is remarkably clear that urban systems such as transportation and water bear an ideology. Just ask anyone who used to live in the black communities that were destroyed last century to make way for highways that connect cities to white suburbs.[18] Or the majority black and impoverished residents of Flint, Michigan, who were poisoned with lead after state officials decided in 2014 to save money by changing the city's water source.[19] Winner famously describes how Robert Moses designed the overpasses on Long Island to be abnormally low as a way to prevent poor and minority New Yorkers (who mostly traveled by bus rather than private car) from reaching his prized beaches.[20]

But the mirage of objectivity is a common fallacy when quantitative and technical methods are involved. "A decision made by the numbers . . . has at least the appearance of being fair and impersonal," explains the historian Theodore Porter. "Quantification is a way of making decisions without seeming to decide."[21]

This siren song of finding objective, technical solutions to social issues is dangerous, especially when we are dealing with technologies as potent as those in the smart city. Believing that such answers exist leads us to underappreciate technology's social and political impacts and ignore alternative approaches for addressing those same issues. By blocking off legitimate political debate in the name of technological progress, presumptions of neutrality tend to bolster the status quo and obstruct more systemic reforms.

This book aims to expose the politics underlying smart cities and shed light on the myriad ways that technology impacts urban governance and life. Smart city rhetoric implies that technology follows an inevitable path, can take only one particular form, and is the primary driver of social and political progress—a common attitude, known as "technological determinism." Tech goggles suggest that adopting newer, faster, and more sophisticated technology is the sole path to improving cities. Instead of questioning how technology should be designed and what social outcomes it should support, technologists present us with the smart city as the only available and attractive urban future.

Technology does not take some inevitable path, however. We shape technology by embedding values in its design and developing it to achieve particular outcomes. Allowing society to be structured by technology thus grants a subtle but potent power to those who design and deploy that technology; we must be critical about the values embedded in these tools and who gets to choose them. Many technologies are designed to remedy social issues by enhancing efficiency, for instance, but that approach does not make them value-neutral. Efficiency is a normative goal: it favors particular principles and outcomes at the expense of others, typically altering how status and resources are distributed across society. Determining which principles should be paramount in enhancing efficiency—in other words, determining *what* should be made efficient, the very question that Cisco dismisses as already resolved—thus requires the inherently political task of mediating between competing normative visions. As the philosopher Marshall Berman implores, "[W]hen we encounter categories like success/failure . . . , we need to ask: By what criteria? By whose criteria? For what purposes? In whose interests?"[22]

In the traffic optimization example discussed above, the efficient flow of autonomous vehicles could mean that pedestrians and cyclists are marginalized on city streets, since their presence would hinder traffic. Similarly, an emphasis on efficiency in civic engagement may position city governments as little more than customer service agencies, compounding inequality by prioritizing relatively superficial civic needs over more substantial ones. As we explore these and other smart city projects, we will repeatedly see how efficiency carries the veneer of being objective and socially optimal but can actually generate unexpected and unjust impacts.

We also project social and political values onto technology by choosing what practices and priorities to support with the capabilities it provides. As technology gets embedded within social and political institutions, its impacts are shaped by the values and practices therein. For example, even if predictive policing algorithms could make accurate and unbiased crime forecasts, that capability still would not dictate how we use those predictions. Our choice to send police to supposed crime hot spots condemns these algorithms to exacerbating the discriminatory practices and policies of our current criminal justice system, regardless of their technical characteristics. But that choice, and hence those impacts of crime predictions, is not inevitable: some cities are using similar algorithms to advance social justice by identifying individuals at risk of incarceration and proactively providing them with social services to keep them out of jail. No matter how advanced our technology may be, in other words, we can never escape from the normative and political task of deciding how to use it.

In addition, a technology's design and political structure can generate social impacts that have little to do with its nominal function. What makes LinkNYC troubling is not the stated application of the technology—free public Wi-Fi is a service that every city should provide—but the way in which that service is achieved: it is funded by collecting and monetizing data about the public. Similarly, local governments are increasingly making important decisions (such as sentencing criminal defendants and assigning students to schools) based on the outputs of algorithms. Yet despite the potentially life-altering decisions that these algorithms inform, cities typically provide the public with little or no insight regarding how they were developed or how they work. Even if these algorithms can improve the accuracy of certain decisions, they contribute to the creation of unaccountable black-box cities. These same technical capabilities can be realized through far more democratic architectures, however, if only we can muster the political will to build them.

Finally, there are many nontechnical factors that constrain the impacts of technology and prevent it from generating outcomes that appear through tech goggles to be predetermined. Many believe that technological advancements in communication will support a bright new era of political engagement and dialogue, for example. But these dreams have not been realized, because the fundamental limitations on democratic decision making and civic engagement are not informational or conversational inefficiencies but rather power, politics, and public motivation. Similarly, even potentially valuable technologies may have underwhelming impacts if they are not properly used or managed. Contrary to the fables told by smart city proponents, technology creates little value on its own—it must be thoughtfully embedded within municipal governance structures.

This book will contain many examples of failures by technology evangelists to accurately forecast technology's impacts—failures caused by their overlooking many of the determinants of social and political issues. While technology can certainly alter social and political conditions, it is also contingent on them: indeed, technology's impacts are largely shaped by the contexts and manner in which it is deployed.

Technological determinism thus warps debates about technology in cities by blinding us to the full range of outcomes that technology could support. When we grant agency to technology, we rid ourselves of the agency to develop visions for the world that we want to create. Presuming a single path for technological development instead leads us into meaningless for-or-against debates in which those who adopt new technology are seen as innovative while those who do not are branded as Luddites.

In this way, the smart city achieves much of its appeal via its juxtaposition to a boogeyman: the "dumb city," a municipality that stubbornly refuses new technology and clings to obsolete and inefficient practices. Acolytes present smart city solutions as a necessary improvement to the dumb city without analyzing technology's social impacts or considering alternative designs. The smart city is thus founded on a false dichotomy and blinds us to the broader possibilities of technology and social change. We become stuck asking a meaningless, tautological question—is a smart city preferable to a dumb city?—instead of debating a more fundamental one: does the smart city represent the urban future that best fosters democracy, justice, and equity?

* * *

I believe that the answer is no—that our essential task is to defy the logic of tech goggles and recognize our agency to pursue an alternative vision: the "Smart Enough City." It is a city free from the influence of tech goggles, a city where technology is embraced as a powerful tool to address the needs of urban residents, in conjunction with other forms of innovation and social change, but is not valued for its own sake or viewed as a panacea. Rather than seeing the city as something to optimize, those who embrace the Smart Enough City place their policy goals at the forefront and, recognizing the complexity of people and institutions, think holistically about how to better meet their needs.

As we interrogate the smart city, we will also read inspiring stories of Smart Enough Cities that are leveraging technology to create lasting benefits for residents. The leaders who spearheaded these efforts did not unlock a special new app or algorithm. Instead of trying to be "smart" and blindly chasing efficiency and connectivity, these leaders realized that cities need to be only "smart enough" to advance their social policy goals.

Several common attributes will emerge among the Smart Enough Cities we explore. The first is that the most impactful applications of technology occur when it is deployed in conjunction with other forms of innovation. In smart cities, technology is deployed to make existing processes and programs more efficient, with little or no critical assessment of how well those processes and programs meet the needs of urban residents; improving cities means improving their technology. In contrast, Smart Enough Cities recognize that social problems are rooted in more than just technological limitations and embrace a variety of approaches (including, but not limited to, technology) to ameliorate those problems.

In the stories we will read about Smart Enough Cities, new programs and policies that thoughtfully reform existing practices generate the main benefits; technology acts as a critical tool to enhance these new approaches, but would have few benefits without them. For example, we will examine how Seattle improved homeless services by restructuring its contracts with social service providers and more clearly defining its goals. Although the city also gained useful data that informs how it deploys resources, its most important innovation was developing a new approach to working with local organizations. Paired together, the data and contract reforms generated impacts far greater than either could have achieved on its own.

The other essential attribute of Smart Enough Cities is that they unlock technology's value by supporting its adoption with reforms to institutions and operations. Visions for smart cities tend to presume that technology operates in a vacuum and that the key to success is having the best tool or the most information. In contrast, with a keen awareness of the many nontechnological barriers to using technology in government, Smart Enough Cities recognize that technology will have little impact unless it is thoughtfully embedded into municipal structures and practices. We will read about how Johnson County, Kansas, created data-sharing processes that are vital to keeping individuals suffering from mental illness out of jail, how New York City and San Francisco developed quality standards and training to help city staff use data, and how Chicago and Seattle developed governance structures to ensure that new technology is used responsibly without violating individual privacy. Tech goggles–inspired headlines about these initiatives might proclaim the power of technology, when in reality the impact of that technology relies on a great deal of bureaucratic (and decidedly unsexy) innovation and support.

The notion of "smart enough" may strike some as shifting the goalposts and setting our sights too low—settling to be merely "good enough." But in fact, the principles of a Smart Enough City are far more ambitious and harder to achieve than those of a smart city. Compared to addressing intractable urban social and political challenges, solving purely technical problems such as predicting crime and deploying Wi-Fi is trivial. Thoughtfully blending technical and nontechnical perspectives to promote democratic and egalitarian cities is the greatest aspiration of all, one that we must ceaselessly pursue.

* * *

This book is about the battle for the future of cities. The smart city may represent the next major urban transformation, with digital technology playing the role today that trains, electricity, and cars have played in the past. But the revolution ahead is not primarily a technological one: as we will see, many smart city technologies fall far short of achieving their promised benefits. We must instead interrogate the smart city because, through the technologies we deploy, we are going to answer some of the most fundamental social and political questions about twenty-first-century cities: Whose needs should urban design prioritize? What is a desirable relationship between a government and its constituents? How should society address crime? How much autonomy should individuals have in their relations with governments and companies? To the extent that the smart city revolutionizes urban life, in other words, it will be by transforming the landscape of urban politics and power than rather than by creating any sort of technological utopia.

With this in mind, we will journey through city halls, tech companies, police departments, and urban neighborhoods to uncover the risks posed by smart cities—and discover why an alternative approach is both necessary and possible. We will examine city governments that have followed the principles of Smart Enough Cities to support policies and programs that improve the well-being of their residents. By juxtaposing successful and unsuccessful applications of technology in cities, this book will identify strategies for ameliorating urban problems using technology as well as strategies for avoiding ineffective and perverse uses of technology.

While the smart city touches many sectors around the world, my main focus will be on how municipal governments within the United States deploy and manage new technology. The reason for this emphasis is twofold. First, those are the bounds of my own competency. I have previously worked for the city government in New Haven and Boston (and have worked closely with other municipalities, including Memphis, San

Francisco, and Seattle) advising on best practices for how to adopt, manage, and use technology. Although I will occasionally look abroad for lessons and parallel developments (indeed, our first story will come from Toronto), my experience is limited to the particular legal and policy environments of U.S. cities.

Second, local governments are taking on an outsized and new role in dictating the social outcomes generated by new technology, creating an urgent need for in-depth analysis of how municipalities use and control technology. City governments are responsible for many of the most impactful decisions about how to deploy new technologies. This is unfamiliar territory for most municipalities, yet it is imperative that they make the right decisions now, while our notions of urban technology are still developing. The decisions we make today will dictate the social and political conditions of the next century. And as the population becomes increasingly urban, it is more important than ever that we thoughtfully assess our hopes and plans for cities.

This book, however, is intended not just for city officials but for all urban dwellers. They are the ones who will reap the benefits and suffer the detriments from new technology—and who must hold their local governments accountable for advancing equitable urban progress with technology. I hope that the lessons this book provides will extend beyond the bounds that constrain my focus and prove useful for activists, technologists, and governments around the world.

The book's structure mirrors the transformation that is needed in our vision for the future of urbanism. This first chapter's theme—the smart city—is today the predominant dream and hence our point of departure. In each subsequent chapter, we will encounter alternative visions for the city that are in conflict with the smart city but nevertheless attainable with the help of technology: a livable city, a democratic city, a just city, a responsible city, and an innovative city. Along the way, each chapter will develop a progressively deeper portrait of how technology impacts society and why cities need to focus on policy, institutions, and people even as they pursue new technology. Together, these stories will demonstrate why cities must strive to be "smart enough" rather than "smart," thereby repositioning technology as a means to improving cities rather than as an end in and of itself. We will conclude by condensing the many lessons learned into a new and bold vision: the Smart Enough City.

The fundamental question of this book is not whether to be for or against innovation, nor for or against technology—it is how to facilitate the innovation and progress that will most benefit city residents. One can oppose a particular implementation of technology without opposing the development and adoption of new technology in general. For there is more to progress than adopting new technology. Progress ought to also mean adapting policies and practices to achieve a more inclusive and democratic city. Designed thoughtfully, technology can be an incredibly potent tool to advance such progress; designed carelessly or inappropriately, technology can inhibit or even derail it.

In this respect, my challenge to the smart city is fundamentally pro-technology: I believe strongly in technology's ability to improve municipal governance and urban life. In fact, it is precisely because of this optimism—because I see how far we risk falling short of an attainable and more desirable urban future—that I write with such urgency.

For technology to achieve the positive impacts of which it is capable, we must dismiss naive dreams of being smart and instead incorporate technology into holistic social and political visions. We must take off our tech goggles and proclaim that smart cities are not what we need. In fact, they distract us from the cities that we do need: livable, democratic, just, responsible, and innovative ones. Technology can help us transform these cities from dreams into reality—but only once we reckon with the critical question that few have asked of smart cities: smart enough for what?

2   The Livable City: The Limits and Dangers of New Technology

In 2014, the first time Steve Buckley saw a picture of Google's self-driving car, he nervously considered the future.[1]

Throughout his life, Buckley had been "infatuated with how easy it was to move about the country." By the time he was thirteen years old, he had already visited forty-nine states. Buckley followed his passion for travel to a career in transportation services—first designing highways in Pennsylvania and Maryland, then serving as deputy commissioner for transportation in Philadelphia, and now holding the position of general manager of transportation services for the City of Toronto. And although Buckley had followed the development of autonomous vehicles (AVs) over those years, he "was always skeptical of it." He thought the technical challenges were "insurmountable."

But when Buckley saw that photo of Google's self-driving car, he realized that AVs were headed for Toronto—and every other city—and that they would transform urban life. "Pretty quickly it became apparent that this was more than just a transportation issue," Buckley says. "What if Google comes in at night and dumps 10,000 of these on our streets? What would we do?"

Buckley was no stranger to the disruptive potential of new transportation technology. In 2014, he was assessing how to respond to the on-demand transportation service Uber, which had recently begun operating in Toronto in defiance of local regulations. Like most cities, Toronto had not developed any plans to assess, manage, or regulate Uber. And given the novel ways that Uber utilized technology to provide transportation, the city was not sure what regulations it could, and should, enforce. So as Uber rapidly expanded its operations, Toronto was struggling to catch up.

Ryan Lanyon, Buckley's colleague in Toronto Transportation Services, saw the revolution presented by Uber and realized that "automated vehicles could be a disruptive force at a much broader scale." When AVs are deployed, says Lanyon, "we can't have the same reaction. We really need to get ahead of this."[2]

Buckley and Lanyon formed an Automated Vehicles Working Group in 2016 to educate division heads and city staff about the potential impacts of AVs. They began with a seemingly simple question—What will vehicle automation mean for Toronto?—and found abundant optimism that AVs hold almost endless promise to improve cities.

First off, vehicle automation could dramatically increase motor vehicle safety. In 2015, almost 2.5 million people were injured and over 35,000 people died in motor vehicle crashes in the United States. Human error is responsible for 94 percent of automobile accidents: in 2015, almost a third of all traffic fatalities were related to drunk driving; another 10 percent of fatalities were due to distracted driving.[3] Self-driving cars—which cannot get drunk, distracted, or tired—promise to eliminate the grim dangers of driving. An analysis from the Eno Center for Transportation found that if 90 percent of cars in the United States were autonomous, there would be 4.2 million fewer crashes per year and 21,700 fewer traffic fatalities—equivalent to saving 60 lives every single day.[4]

Self-driving cars could also rapidly increase the speed of travel. With improved perception, connectivity, and reactivity in comparison with human drivers, AVs may be able to travel at high speeds without the need for large headways. The Eno Center estimated that at 90 percent AV penetration, roadway capacity would double and congestion would fall by up to 60 percent.[5] The chief technology officer of a major automotive parts manufacturer has proclaimed that "if every car was talking to each other, traffic flow would be incredibly smooth: no traffic jams."[6] So drastically will AVs relieve congestion that the urban designer Kinder Baumgardner has named them "clairvoyant vehicles."[7]

AVs' increased perception and their heightened ability to communicate with urban infrastructure might even enable cities to remove that indelible signifier of urban traffic: red lights. "Imagine a city without traffic lights, where lanes of cars merge harmoniously from one to the next, allowing traffic to flow smoothly across intersections. This futuristic vision is becoming reality," proclaims the Senseable City Lab at MIT.[8] By replacing traditional intersections—"natural bottlenecks"—with intelligent intersections that act as "orchestra conductors" for a city's vehicles, the MIT researchers suggest, cities could double street capacity and significantly reduce traffic delays.[9]

AVs could also enable transformations in urban design. "As speeds increase, fewer lanes are needed and more freeway lanes will be decommissioned," predicts Baumgardner.[10] There is even more excitement about AVs drastically reducing the need for parking in cities. As cars begin to drive themselves, it may no longer be necessary to leave cars sitting at downtown curbs or parking lots all day. Instead, a self-driving car could drop off a passenger in front of her office and then zoom away to pick up another passenger or park itself in an out-of-the-way location. According to the manager of Audi's Urban Futures Initiatives, "Parking will be moved indoors and outside of city centers, freeing up outdoor lots and spaces for development and public space."[11]

AVs will provide further benefits by relieving people of the burden of driving, promising mobility for broad swaths of the population who currently lack access to transportation or the ability to drive. A driver's license may no longer be a mobility barrier for the elderly, the disabled, and children. For example, a 2017 report concluded that autonomous vehicles could help 2 million people with disabilities access jobs and 4.3 million access medical appointments.[12] Moreover, because people will no longer be responsible for driving, time spent in a car may be reclaimed for other purposes. Morning commutes could become opportunities to catch up on email, read the news, or watch TV.

Given all of these expected benefits, there is remarkable optimism about how autonomous vehicles will improve cities. In 2013, Morgan Stanley reported that a world of ubiquitous AVs is possible by 2030 and will be a "utopian society."[13]

This may sound amazing, but we have heard such promises before. The Motor Age envisioned in the 1930s was to be "an automotive millennium without accidents, congestion, or delays."[14] The 1939 World's Fair in New York City highlighted a General Motors–sponsored "wonderland" called Futurama, which prophesied a "modern traffic system" with "greater efficiency" that would enable the "elimination of congestion" and "better ways of living."[15]

When it comes to urban transportation, we have been waiting for a century for new technology to provide a solution. And just as last century's cities were re-created in the image of Futurama, leading to congested high-ways and meager pedestrian and public transit facilities, so today's cities may become landscapes that serve, above all else, the optimized flow of self-driving cars.

As they considered the potential impacts of AVs in Toronto, Buckley and Lanyon grappled with a seemingly endless tangle of questions and scenarios. Indeed, every city faces the daunting task of determining how it will be affected by the cars of future. But the most important task for a city is not to predict the future of technology and hope for the best—it is to shape its own future through thoughtful use of technology.

* * *

To avoid repeating history—to avoid falling into the Futurama trap—we must learn from the past. The process by which the Motor Age came to be desired and pursued demonstrates the dangers of putting too much faith in technology to solve social problems and highlights the critical decisions we must make today about how to prepare for self-driving cars.

At the start of the twentieth century, it was commonly accepted that streets were public spaces where streetcars could run, people could walk, and children could play. When cars were introduced in large numbers onto the streets of American cities in the 1920s, they brought chaos and conflict. Gruesome accidents horrified the public. Parents feared for the safety of their children. Downtown business owners worried that congestion would diminish profits. Early attempts by police officers to create order in the streets proved fruitless. There seemed to be no way for cars to peacefully coexist with pedestrians, children, and streetcars.

The car was an "intruder" in the existing balance of city streets, writes the historian Peter Norton in *Fighting Traffic*. As a new technology "incompatible with old street uses," Norton explains, cars "violated prevailing notions of what a street is for."[16] The resulting destabilization instituted a period of "interpretive flexibility," during which social conceptions of cars and streets were in flux.[17] Motorists, families, police, businessmen, and automotive manufacturers all jockeyed to define how cars should be used and who had a rightful claim to the street.

In need of a neutral way to mediate between these parties, cities turned to engineers for a solution. Despite the contentious nature of managing urban streets, traffic engineers were trusted as "disinterested experts" to solve the problem.[18] Because they "made [deductions] in a scientific manner,"[19] it was commonly believed that engineers could devise an objective and socially optimal solution.

Over the previous several decades, engineers had displayed technical expertise in helping cities efficiently manage overburdened public utilities such as water and electricity. There was no reason to think that traffic was any different. To engineers, writes Norton, "City streets were . . . like water supply, sewers, or gas lines: a public service to be regulated by experts in the public interest."[20] Engineers derived new methods (such as the traffic survey) from their work managing these other utilities. Likening the flow of traffic to the flow of water or sewage,[21] they were confident that "scientific organization of traffic . . . could cut traffic congestion at once by half."[22]

Here we see our first instance of tech goggles, a prelude of what is to come: in keeping with their past work on other public utilities, traffic engineers "stood for the logic that efficiency worked for the benefit of all," writes Norton. "They saw their mission as optimizing traffic capacity."[23] And so engineers began altering traffic signal timings based on equations developed to maximize how many cars each street could carry.

Improving traffic flow came at a cost, however, for optimizing one aspect of urban life requires restricting others that would impede that efficiency. Motorists may have benefited from the updated signal timings that enabled the faster flow of cars, but pedestrians discovered that these changes had made streets inhospitable. Navigating city streets became, in the words of a 1926 *Chicago Tribune* report, a "succession of heart thrills, dodges, and jumps."[24]

By focusing on vehicle speeds and ignoring the needs and behaviors of pedestrians—who were left out of the equations entirely—traffic engineers increased traffic flow but, in the process, Norton explains, "helped to redefine streets as motor thoroughfares where pedestrians did not belong." In turn, streets became "socially reconstructed as places where motorists unquestionably belonged."[25] Through a process known as "closure," the interpretive flexibility about streets created by the introduction of cars yielded to a social consensus that streets were for vehicles and that pedestrians who got in the way were troublesome "jaywalkers."

This shift in social conceptions prepared the way for the auto industry to promote self-serving arguments that cities should be redesigned to prioritize and facilitate the passage of cars. Congestion was blamed not on the spatial inefficiency of cars but on insufficient street space. Similarly, the dangers of cars were framed as failures of pedestrians and antiquated streets. Through advertising campaigns and scale models such as Futurama, automobile manufacturers, oil companies, and others with a financial stake in the growth of cars and highways generated popular support for a utopian Motor Age in which cities would be remade for cars.[26] These groups leveraged their newfound muscle into massive government investments, most

Figure 2.1
The Futurama exhibit created by Norman Bel Geddes for the 1939 World's Fair in New York City. Futurama, which was sponsored by General Motors, portrayed a vision for the city of 1960, where automobile collisions and traffic congestion would be eliminated.
Source: Norman Bel Geddes, *Magic Motorways* (New York: Random House, 1940), p. 240. Copyright © The Edith Lutyens and Norman Bel Geddes Foundation, Inc.

notably the Interstate Highway System: authorized in 1956, it represented the largest domestic public works program ever undertaken at that point in world history.[27] It is only in the past several decades, through a renewed focus on designing cities for people rather than cars, that many of these automobile-focused designs have been reversed.

Embedded in the focus on efficient car travel are two critical flaws, both of which rear their head today as we begin planning for self-driving cars. The first issue with traffic models, as just described, lies in what they choose to measure and ignore. While engineers go to great lengths to evaluate automobile flow, they pay far less attention to the throughput and safety of pedestrians and cyclists, as well as to public transit. "When a traffic engineer says they've optimized a traffic signal, that typically means they made it the best for the motorists," explains one transportation engineer.[28] Another notes, "*Synchro*, the standard software [that traffic engineers] use, is based on minimizing auto delay and it doesn't even calculate pedestrian delay."[29]

Because most traffic engineers strive for efficiency—defined solely in terms of automobile travel—they do not measure whether roads meet the needs of pedestrians and transit riders. And what is left out of equations tends to be not just ignored but devalued. Stripping away pedestrian facilities eases traffic, while the costs to pedestrians, cyclists, and others are not apparent in the models; such action therefore appears, quantitatively and scientifically, to be an unequivocal boon for society. In turn, engineers devise solutions for urban congestion without considering their full impacts on people and communities.

Although an explicit plan to turn streets over to cars would have been met with strong resistance, the use of mathematical models to improve the efficiency of urban streets masked this radical transformation under a veneer of objectivity. There was little recognition that increasing traffic efficiency could benefit certain groups at the expense of others.

The second major flaw with traffic models can be best elucidated with another history lesson. In 1936, New York City opened the Grand Central, Interborough, and Laurelton parkways to great fanfare. After years of heavy traffic congestion throughout New York, these ambitious new projects dreamed up by the "master builder" Robert Moses promised to solve the region's traffic woes "for generations." But instead of generations, traffic relief lasted just three weeks, reports Robert Caro in *The Power Broker*, his epic biography of Moses.[30] Undeterred, Moses continued to build. The Triborough Bridge opened in 1936; the Wantagh State Parkway Extension, in 1938; and the Bronx-Whitestone Bridge, in 1939. Every time, traffic relief was promised. Every time, congestion remained severe.

Planners began to notice a perplexing pattern: "every time a new parkway was built, it quickly became jammed with traffic, but the load on the old parkways was not significantly relieved."[31] Cars seemed to appear out of nowhere. Congestion was so bad following the opening of the Triborough Bridge that the *Herald Tribune* exaggeratedly described observing a "cross-country traffic jam." As the newspaper put it, "motoring residents of the Bronx . . . decided at the same moment to head for the ocean by way of the new bridge and the Grand Central Parkway. And nearly all of them got stuck—as did countless other motorists."[32] In ways that New York's planners and engineers "did not even pretend to understand," recounts Caro, "the construction of this bridge, the most gigantic and modern traffic-sorting and conveying machine in the world, had . . . failed to cure the traffic problem it was supposed to solve."[33]

Moses's New York may represent an extreme example—as so many people were eager to travel that new roads filled almost instantly—but it is indicative of a common phenomenon known as "induced demand." The economist Anthony Downs first defined induced demand in 1962, when he determined that on "urban commuter expressways, peak-hour traffic congestion rises to meet maximum capacity."[34] Downs pointed to several factors in explaining why, "if a road is part of a larger transportation network within a region, peak-hour congestion cannot be eliminated for long on a congested road by expanding that road's capacity."[35] The most obvious reason is that drivers who previously had taken other routes start traveling on the expanded, faster road (Downs calls this "spatial convergence"). Meanwhile, drivers who formerly scheduled their travel to avoid congested periods take advantage of the increased road capacity and start driving during peak hours ("time convergence"). Others stop taking public transportation and start driving ("modal convergence").[36] Additional causes of induced demand include people making discretionary trips they would have forgone in the face of too much congestion and the increased travel needs that sprawling development (facilitated by increased road capacity) creates.

Recent analyses have corroborated Downs's observations. In a 2011 study of urban traffic patterns between 1983 and 2003, the economists Gilles

Duranton and Matthew Turner determined that increasing road capacity generated proportional increases in driving. They conclude, "Our results strongly support the hypothesis that roads cause traffic."[37]

Drawing on their work with public utilities, last century's traffic engineers had mistakenly assumed that cities contain a relatively fixed quantity of transportation needs, so that increasing road capacity would enable everyone to reach their destinations more quickly. But in fact, the primary circumstance keeping many people off the road is congestion. Increasing roadway capacity invites more people to take trips that they would have forgone to avoid traffic. Because engineers overlooked how new or expanded roads would change behavior, they failed to incorporate this second-order effect into their mathematical models and thus drastically overestimated the benefits of increasing roadway capacity. More people might be able to get around and at faster speeds, but congestion is far from eliminated.

* * *

Technology enthusiasts are repeating mistakes made in the past when they diagnose self-driving cars as the route to a "utopian society." They ignore the multiplicity of needs in cities and the complexities of traffic, instead devising narrow solutions that revolve around technology. In fact, the utopian society that features ubiquitous self-driving cars is incoherent and undesirable.

Any realistic prognostications for AVs and traffic must begin with induced demand. By increasing travel speeds and the density of vehicles on the road, introducing autonomous vehicles to city streets is largely equivalent to expanding the physical capacity of those streets. And given that travel demand rises when roadway capacity increases, people will take advantage of these benefits by driving more. This induced driving will add congestion, especially during peak commuting hours, largely negating the benefits that faster travel might provide.

The phenomenon of induced demand also suggests that autonomous vehicles will generate even more sprawling urban development than is already typical in the United States. Counterintuitively, although average travel speeds have risen notably over the past century, average travel times have remained remarkably consistent, because travel distances have also increased: research has shown that people take advantage of increased travel speeds not by enjoying shorter commutes but by moving further away from the urban core.[38] We should thus expect that to the extent AVs enable faster travel, they will lead to increasingly spread-out communities rather than shortened commutes.

Moreover, if time that was previously spent driving is reclaimed for work or leisure, people may be willing to accept even longer travel times, further increasing the distances traveled. Such AV-enabled sprawl could lead to disinvestment in downtowns while also having devastating consequences for the environment: the further away people live, the more they drive and the more greenhouse gases their vehicles emit.

Similar logic explains the challenges of reclaiming parking infrastructure for pedestrian plazas, bike lanes, and apartment buildings. After dropping off passengers, self-driving cars could drive themselves to pick up other passengers or park in peripheral locations, unlocking valuable downtown real estate for alternative purposes. But just because the cars are empty does not mean they are not on the road. If parking lots are developed outside of city centers, self-driving cars will need to get to and from those facilities. If self-driving cars take frequent zero-occupancy trips in and out of the urban core, the number of vehicles on the road could increase dramatically. Instead of being congested with people circling around to find parking, cities may become congested with empty cars driving in and out of downtown. Alternatively, if congestion is too severe, many people may consider it cheaper or more convenient to leave cars in traditional downtown parking facilities. This choice would severely hinder efforts to refashion existing parking infrastructure for more productive uses.

Most importantly, dreams of autonomous vehicles also repeat the mistake of prioritizing traffic efficiency over walkability and community vitality. Consider the claim that travel times will be severely reduced by increased travel speeds and the elimination of red lights. Sounds fantastic—if you're in a car. What kind of city would this create for everyone else? The MIT simulations that demonstrate cities without traffic lights show cars traveling seamlessly through an intersection with remarkable efficiency when compared with their movement on traditional streets.[39] But there's one important element missing: people. The simulations do not include a single person walking, cycling, or riding a bus. Yet the intersection shown is among the most walkable locations in the entire United States[40] and is crossed by some of Boston's busiest pedestrian and transit corridors. If even

A

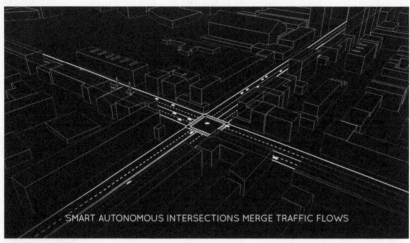

SMART AUTONOMOUS INTERSECTIONS MERGE TRAFFIC FLOWS

B

Figure 2.2

(a) A screenshot from the MIT Senseable City Lab's demonstration video of a city without traffic lights, depicting autonomous vehicles zooming through an intersection in downtown Boston without needing to slow down.

(b) A photograph, taken on a typical Saturday afternoon, of the same intersection, where cars shared the street with pedestrians, cyclists, and buses.

Sources: (a) Senseable City Lab, "DriveWAVE by MIT SENSEable City Lab" (2015). http://senseable.mit.edu/wave/. (b) Photograph by Ben Green. Boston, Massachusetts. April 2018.

this location has been turned into a high-speed interchange, it is difficult to fathom where everyone in those cars is in such a rush to go.

If we want cities in which people are able to cross the street—a reasonable desire, one would think—then we must avoid visions of downtown intersections where AVs speed through without ever stopping. Even if we permit occasional red lights to allow people to cross the street, allowing high speeds on city streets would severely diminish safety, walkability, and vitality. Main Streets would turn into high-speed corridors: imagine how unpleasant it would be to eat lunch or go shopping along the side of the freeway. Although the prospect of unencumbered AV travel is exciting to many technologists, it is not a central feature of successful urbanism. A city devoid of traffic lights in the interest of enabling high-speed streets would also be devoid of people and character.

In conceiving of traffic as an optimization problem that requires a technical solution, proposals for AV-filled smart cities remove all normative concerns from consideration and position traffic efficiency as a neutral and socially optimal objective for cities. Although enabling cars to travel more efficiently is valuable, it is not the only priority for cities. More importantly, efficiency involves political calculations: What should be made efficient? Who gets to decide? By what means should efficiency be attained?

The answers to these questions can have enormous social and political consequences. What society chooses to measure and optimize is an embodiment of our priorities. So long as we value smooth car traffic over livable streets and public transit, efforts to enhance transportation will actually be aimed at easing congestion. Just as last century's efforts to make streets more efficient for cars prompted radical urban designs that benefited automobiles over pedestrians and streetcars, modern attempts to facilitate efficient travel for self-driving cars could prompt urban designs that benefit AVs (and their passengers) at the expense of pedestrians, transit, and public space. At the very moment that cities are undoing the damage created by last century's misguided dreams, we appear ready to revert back to our bad habits.

Framing traffic as a technical challenge also provides cover for private companies to promote their corporate agenda under the supposedly neutral premise of improving efficiency, much as the auto industry did with cars and highways last century. These days, Ford is promising that self-driving cars will create "a future where traffic congestion is drastically reduced."[41] Lyft has taken this argument even further, claiming that "end[ing] traffic

[is] really simple."[42] Its co-founder John Zimmer has said that with AVs and technology that encourages people to carpool, "for the first time in human history, we have the tools to create a perfectly efficient transportation network."[43]

Not only do such proposals falsely present intractable mobility and congestion challenges as easy to solve with new technology, they also blind us to how other approaches—in this case, alternative transportation modes and urban planning policies—can more effectively address these problems. Instead, these corporate pitches focus on how to make driving in cars more efficient, inevitably suggesting that their product or service is the solution. And because these scenarios are presented as optimizing traffic efficiency, they can be framed as advancing universally desired outcomes rather than profits. In response to this rhetoric, some cities and states have considered reducing investments in public transit, expecting that self-driving cars will make such systems obsolete,[44] and begun rolling out the red carpet for AV companies. No state has reduced AV regulations more aggressively than Arizona, leading several companies to flock to Phoenix[45]—and leading also to the first recorded pedestrian fatality from a self-driving car, in Tempe in March 2018.[46]

These limits and caveats do not entirely negate the benefits of autonomous vehicles, but they highlight where dreams involving supposedly beneficial technologies can go wrong, underscoring the gravity of decisions that cities must make in the coming years. Self-driving cars will almost surely enhance safety and mobility, and in some cases might even enable parking facilities to be reimagined for new purposes. There is much to look forward to. But AVs will not create utopias—and the best way to avoid creating dystopian futures is to recognize the limits of self-driving cars and the barriers to their successful implementation.

It is easy to imagine how developing our expectations of AVs under the influence of tech goggles could lead, through the tech goggles cycle, to cities that are optimized for self-driving cars but unwelcoming to pedestrians, transit, and vibrant public space. First, as we have seen, tech goggles cause many to perceive improving urban mobility solely in terms of making traffic more efficient—getting every car from its origin to its destination as quickly as possible. These beliefs lead technologists and cities to prioritize AVs as the solution to almost every transportation issue. As cities are designed to facilitate self-driving cars and, in turn, alternative transportation modes are neglected, people will have little choice but to rely on AVs for mobility, and we may become even more blind to other priorities and potential solutions.

These dangers animate the distaste of the urban planner and *Walkable City* author Jeff Speck for attempts to optimize traffic flow. "What upsets me most about traffic studies is the hegemonic dominance that they hold in the municipal discourse," he explains. "Somewhere along the line, we decided as a society that the only inviolable principle in the design of our communities was that we had to fight traffic congestion. Shouldn't the questions be, Will it increase vitality? Will it increase equity? Will it increase the success of our city?"[47] In this light, Speck believes that "autonomous vehicles are the right answer to the wrong question."[48]

* * *

With an inquisitive yet principled approach to automated vehicles, Toronto is demonstrating how a Smart Enough City can consider the right questions and priorities when preparing for the changes promised by revolutionary new technologies.

Reflecting on the potential and limits of self-driving cars, Steve Buckley realized that Toronto needed to actively pursue the future it desired rather than passively allow technology to dictate the city's future and hope for the best. "Why are we just letting it happen to us?" Buckley asked himself. He shifted the conversations of the Automated Vehicles Working Group away from their initial focus—"What will vehicle automation mean for Toronto?"—and toward a different question: "How do we plan for AVs and how do we shape it?" Because, as Buckley explains it, "We can't let the technological tail wag the city dog."

After all, as Ryan Lanyon observes, self-driving cars come with opportunities but are no silver bullet. "We know from past experience that any efficiencies gained can induce additional demand," he says. "Whether vehicles are automated or manually driven, there's still a limit to the number that can be accommodated in our existing road space. Mass transit is still the best and most efficient way for our municipality to move large quantities of people in certain areas and certain corridors."

Buckley and Lanyon began considering how Toronto could prepare for and shape the forms that self-driving cars take. The key questions, according to Buckley, are, "What are the positive aspects of these systems, and what are the potential negatives or downsides? How do you structure these systems in a way that permits you to get as many positives out of them as you can, and to future-proof yourself against problems?"

Toronto, Lanyon adds, "has a vision for what it wants to become. It wants to be more equitable. It wants to be more sustainable. It wants to continue to develop economically." Lanyon points to Toronto's efforts over the past decade to invest in transit and walkability, rather than developing new facilities or increasing capacity for automobiles. He explains, "We want to reduce congestion. We want to encourage people to use public transit and active transportation. We want to make it a more livable city. We want our streets to be more attractive. Whether we have vehicle automation or manually driven vehicles, we have the same objectives." To Lanyon, the essential question is not how Toronto can optimize itself for AVs, but rather, "How do we harness disruption and paradigm shift to get us toward the goals that are already established?"

To answer this question, Lanyon and Buckley began leading the Automated Vehicles Working Group through assessments of how AVs could advance Toronto's priorities. They analyzed the benefits and drawbacks of the potential ownership models for self-driving cars: private ownership (much as cars are owned today) and shared, on-demand use (like Uber). While both scenarios provided safety benefits, Buckley explains that "a shared model would be better than having a continuation of private ownership of automated vehicles." On-demand AVs are more likely to reduce the need for downtown parking, shrink the number of cars on the road, and expand mobility for many who cannot afford to own a car. Private ownership, on the other hand, might enhance roadway capacity but would also generate longer trips, lead to more unoccupied vehicles on the street, and contribute to increased sprawl.[49]

The Working Group also began planning for a future with AVs with more of an eye to infrastructure, so that no matter how self-driving cars come to be owned, Toronto can take advantage of the technology's benefits. To make roads safer for AVs, the city is assessing the need for improved pavement markings and is researching traffic lights that wirelessly broadcast their signal rather than require cars to have a direct line of sight. And to aid in reclaiming parking space for new purposes if the opportunity arises, the city is reviewing its parking infrastructure and regulations.

Meanwhile, the AV Working Group continues to facilitate discussions and educate public officials through scenario-planning exercises. It plans to launch a pilot in 2020 using self-driving shuttles to improve accessibility to and from transit stations.[50]

In following this approach, Toronto is transcending the false dichotomy between the smart city and dumb city: rather than unreservedly embracing AVs or rejecting them entirely, the city is sticking to its planning and transportation goals while exploring the opportunities offered by technology to achieve them. In this way, it avoids the smart city trap of embracing technology without considering how to create more livable urban environments. For although an automobile-dominated city with AVs is likely preferable to an automobile-dominated city without AVs, both pale in comparison to a livable city where public transit, walkability, and public space can thrive.

"We have an opportunity to start talking about it and getting it right," Buckley says. "It's better to do that now than to let the genie out of the bottle and try to put it back in."

* * *

Some cities are going even further than Toronto, moving beyond planning exercises by taking bold steps to improve mobility and livability through technology. Leading the way is Columbus, Ohio, winner of the Smart City Challenge that the U.S. Department of Transportation (DOT) launched in December 2015. The challenge was designed to spur mid-sized U.S. cities to plan for a "first-of-its-kind smart transportation system" and came with a $40 million prize[51]—money that one city could use to revolutionize its transportation ecosystem.

Growing up in Columbus, Jordan Davis always had a chip on her shoulder. "Columbus can never stand alone without the 'comma Ohio,'" she laments. "Nobody knows us." A graduate of Ohio State University and the proud daughter of a chamber of commerce president, Davis says that "it's in my DNA to want to build a better city." Given Columbus's recent economic renaissance and downtown revival, not to mention its status as the Midwest's fastest-growing city, Davis has long been eager to bring more attention to everything her hometown has to offer.[52]

When the Smart Cities Challenge was announced, several local organizations joined forces under a new umbrella organization called Smart Columbus, gathering in a joint working space in a local startup incubator to develop a proposal. These efforts were rewarded in June 2016, when the DOT announced that Columbus had beaten seventy-seven other cities (and a group of seven finalists that included Austin, Denver, and San Francisco) to win the challenge. What stood out about Columbus's efforts was not a futuristic plan to end traffic with a fleet of AVs but its focus on addressing transportation hurdles that diminish social welfare.

"This is exactly what Columbus needs," says Davis, who directs Smart City Strategy at the Columbus Partnership, a nonprofit civic organization that emphasizes local economic development. "Transportation has not always been our shining star. I think this is our moment to think differently about what our future can look like."

That Columbus beat numerous other cities, including several viewed as hotbeds of the country's transportation and technology sectors, was no fluke. "It's not like we were sitting here waiting for something like this to happen," explains Thea Walsh, director of transportation for the Mid-Ohio Regional Planning Commission (MORPC), which played an integral role in articulating the city's vision for the challenge. "We had been doing a lot of planning locally, and when that opportunity came out, we realized, 'Wait a second, that kind of sounds like the things we're talking about and planning for.'"[53]

Columbus had spent the previous years identifying and addressing gaps in its transportation system and had begun exploring the opportunities that new technology could provide. "We enjoyed wonderful sprawl in central Ohio from 1880 to 2010," MORPC's planning director Kerstin Carr sarcastically explains. "There is a huge reliance on single-occupancy vehicles," adds her colleague Walsh. "We're not a community that has high-capacity or high-quality transit systems whatsoever. This is very much a bigger city operating like a cow town when it comes to transportation." Given its projection that the region will grow by 500,000 people and 300,000 jobs by 2050 (from a 2010 population of 1.8 million),[54] the city recognized that it needed a new approach.

In 2013, MORPC conducted a long-term planning assessment called insight2050. It considered four potential growth scenarios that the region could pursue over the next several decades, ranging from sprawling development that followed past trends to dense development that maximized infill and redevelopment. Each scenario was evaluated on outcomes related to land consumption, energy use, transportation, and costs.

Across the board, the results were clear: "the more dense, the more mixed-use, the more walkable and compact our communities are getting, the better," says Carr. Compared to the future that would result if Columbus continued following its traditional planning approach, the densest scenario was projected to decrease the region's total driving by 30 percent, reduce greenhouse gas emissions by 33 percent, save the city $80 million per year, and significantly improve public health.[55]

This was "a very low-tech conversation," emphasizes Walsh, "but it needed to be had so that we can facilitate better service." Only after the groundwork of developing a planning vision for the future had been laid does Walsh believe that the city was ready to begin thinking about how technology could advance its goals, making it possible to win the Smart City Challenge.

With the DOT grant in hand, Smart Columbus is experimenting with new approaches to address the issues that insight2050 identified. One of its aims is to improve mobility around Easton, a major office and retail park northeast of downtown Columbus. "It's not accessible," notes Carr. Getting to Easton from the nearest transit stop requires crossing ten lanes of traffic and then walking a while further. Moreover, Easton itself is sprawling and difficult to navigate, forcing people to drive between sections or else remain in isolated pockets. To improve transit accessibility and mobility within the complex, Smart Columbus plans to deploy self-driving shuttles. One will connect the transit center to Easton; another will travel within the complex. Hopefully, says Davis, this solution will help create "an independent mobility experience if you don't have a car."

Beyond improving mobility and transit generally, Columbus's efforts embody a vision to mitigate inequality and improve social welfare. "You have to democratize mobility," explains Carla Bailo, a former Nissan executive who runs mobility research at Ohio State University and who played an integral role in rallying the community to apply for the Smart City Challenge.[56]

Smart Columbus's efforts to address inequality through transportation are focused on Linden, a neighborhood situated between downtown Columbus and Easton that suffers from three times the unemployment rate and has less than half the median income of Columbus as a whole.[57] A critical issue in Linden is a lack of access to prenatal and early childhood healthcare, leading to an infant mortality rate that is more than twice the city average.[58] Because few who live there can afford a car and the city provides only meager public transportation options, Linden residents often miss or are late to doctor's appointments. At one community forum, half of the residents agreed with the statements "I don't take the bus because it takes too long to get to my destination" and "It is too far to walk from home to where I want to go." According to one resident who spoke at the meeting, "there just isn't enough time to do everything in a day and get to where you need to go."[59]

Hearing these issues, most technologists would instinctively prescribe autonomous vehicles as the panacea. Indeed, Bailo reports that Smart Columbus first approached this problem "focused solely on first-mile and last-mile solutions," referring to the challenge of getting people to and from bus stops and transportation hubs. But as the group talked to Linden residents about their needs, Smart Columbus realized that Linden's barriers to jobs and healthcare went deeper than a lack of convenient transportation.

"It wasn't just about moms not getting to the doctor, but also that there was basic information missing," explains Bailo. While most Linden residents have smartphones, many lack access to a data plan or Wi-Fi. And because information about public transportation in Columbus is scattered across multiple websites and apps, even those who do have internet access struggle to determine the best way to get around. On the basis of this research, Smart Columbus is improving Wi-Fi access in Linden, especially at schools and community centers, and creating a streamlined app that unifies every transportation option. "If we ask them to surf all the different websites and create a bunch of different accounts, we're doomed to fail," says Bailo. "If we can have one simple app that provides them with their alternatives and also allows them a common payment in that app, it's doable."

An additional hurdle is that many Linden residents are locked out of transportation apps because they lack access to a bank account or credit card. Uber, Lyft, and the local car-sharing and bike-sharing companies all require a credit or debit card to pay for trips. The local bus system, on the other hand, is cash-only, which makes it difficult for social service providers to subsidize transport to doctor's appointments or work. To address these issues, Smart Columbus is creating a unified payment card and app that will make it possible for users to pay for all of the region's transportation modes. Kiosks at key locations will enable people to load cash into their accounts.

Another challenge that Linden mothers face is getting to doctor's appointments for prenatal or newborn care when they already have young children. Getting to and waiting at a bus stop may simply not be feasible. To help remove these barriers, Smart Columbus is developing a subsidized, on-demand ride service to take pregnant Linden moms directly from their homes to medical appointments and back,[60] while also exploring how improved childcare facilities could make it more feasible for Linden residents to go to job interviews and doctor's appointments.

Columbus demonstrates two attributes key to fostering Smart Enough Cities. First, cities need to have a clear policy agenda before deploying technology. Thinking about the city's challenges and needs before thinking about technology is essential, according to Bailo. "As a city you really need to define what are our problems, how do I prioritize those, and how can technology and data make those better? Then you'll be able to make a road map for your city, based on the problems that exist and the prioritization of those," she explains. "Improving people's lives is the key element. Because otherwise you're just throwing technology and data out for the fun of it."

The second necessary ingredient is a research process that focuses on people rather than technology. As Columbus demonstrates, the best way to avoid the simplistic and solutionist mind-set fostered by tech goggles is to learn what barriers and challenges people actually face. This means getting out and talking to city residents. "What we thought was this basic problem—get people from A to B—turned into an entire support system that needs to be put in place to tackle these issues," observes Bailo. "We really needed to look at it from a more holistic viewpoint and to think about many different ways to provide transportation for the residents in this community. As geeky technology people, we wouldn't have thought about these things had we not considered the whole picture."

Despite these promising efforts, the path ahead remains difficult. Developing plans to aid an underserved community and actually serving its needs when flashier projects can be pursued remain two very different tasks. Some have worried that the focus on prenatal healthcare is waning.[61] Columbus also knows that it must learn from its past urban development mistakes: namely, a century of sprawl and the creation of disconnected neighborhoods that bred poverty. "We don't want to put ourselves in these situations again," says Walsh. "Because if we make the wrong decisions now, we'll be so tightly locked in that it will be hard to come back out of it."

The past also provides hope, however. If Columbus is successful, the city will resume its position, held long ago, as a leader in transportation technology. In 1900, the Columbus Buggy Company (CBC) was the world's biggest buggy manufacturer, producing one-fifth of the world's supply; Columbus was the "Buggy Capital of the World."[62] The CBC also produced some of the first electric vehicles, which could travel 75 miles on a single charge. But when Henry Ford and the Model T came around in 1908, the CBC could not keep up. It went bankrupt in 1913.

Today, as part of the city's downtown revival, the old Columbus Buggy Company warehouse has been remodeled as residential lofts. For Jordan Davis, a former resident of "The Buggy," this local history is a source of constant motivation. "Think of the disruption that happened when we went from horses to cars—we were dramatically impacted by that—and how dramatic it will be when we go from humans driving to machines driving. Hopefully Columbus can win this one."

* * *

The past and future of transportation demonstrate why tech goggles are such a dangerous frame through which to pursue new technology. First, tech goggles blindly identify innovation and progress with technology. In assuming that a complex problem is easy to solve with technology, we overlook more systemic changes that may be necessary. Instead of deliberating about what type of city to create—and how AVs can support those outcomes—we consider only how to make existing cities more efficient with AVs. While automated vehicles clearly promise significant benefits (such as enhanced safety and mobility), just as motorized vehicles introduced important advances a century ago, cities are defined by more than efficient traffic flow. The preoccupation with cars (both self-driving and not) distracts us from the need for holistic urban livability as well as from other strategies to improve mobility: public transit, dense development, and congestion pricing, not to mention user-friendly apps and better childcare.

As we learned with cars, the mere fact that social norms revolve around a particular technology does not imply that the technology is optimal. This is the danger of closure, the process by which we come to a consensus about certain technologies. The historian Thomas Misa explains, "closure occurs . . . not when a neat solution emerges but when a social group perceives that the problem is solved." In fact, Misa adds, "closure may obscure alternatives and hence appear to render the particular artifact . . . as necessary or logical."[63] In this sense, we can conceive of the tech goggles cycle's

"reinforcement" stage as representing a form of closure linked to the particular technological arrangements suggested by tech goggles.

Because closure following the embrace of a suboptimal technology can
blind us to better alternatives and lock us in to harmful practices, we must
reject efforts to design cities for AVs. As automated vehicles begin to appear
on city streets, a new period of interpretive flexibility is emerging. And given
that "this flexibility tends to be greatest when an artifact is new,"[64] the decisions we make over the next several years will shape cities for decades. If
we imagine cities as traffic optimization problems that only AVs can solve,
we are likely to destabilize the emerging consensus that cities should foster
dense, walkable neighborhoods and to instead usher in a new paradigm of
designing cities to accommodate self-driving cars. The more we design society for ubiquitous cars, the harder it will be to pursue alternative visions if
and when we eventually recognize their merits.

The second danger of tech goggles that emerged in this chapter is the
tendency of optimization and efficiency to mask political decisions as
objective, technical ones. When we misinterpret complex social issues as
technology problems, we evaluate solutions along purely technical criteria
and overlook their political consequences. Political debate is reduced to
narrow technocratic discussions of efficiency.

This approach is blind to the full effects of making some aspect of society
more efficient. Although traffic engineers were considered neutral because
they employed a "scientific" approach, their models that optimized for
automobile efficiency and ignored other street users had revolutionary
social impacts. Similarly, today's models that optimize for AVs are presented
as utopian and objective, but their disregard of social impacts beyond
smoother traffic quite literally means that cities will be altered to prioritize
the needs of self-driving cars over those of pedestrians and communities.

Casting social issues as technical challenges also allows companies to
advance their corporate agendas without appearing partisan. Just as last
century's auto industry promoted the "Motor Age" as liberating and universally desirable, so today's technology companies promote the "smart city"
as a scientific way to enhance efficiency and daily life. Our recognition of
the Motor Age's true proponents and the unfortunate results of its pursuit
should make us skeptical about the smart city and the agenda that under-
lies its promises. Allowing tech companies to drive us toward the closure of

cities becoming "smart" repeats the damage of allowing the motor industry to push us toward the closure of cities embracing cars.

Smart Enough Cities must stay true to their priorities and embrace the benefits of new technology without falling prey to technological solutionism. Columbus demonstrates why it is so important to focus on real people and issues in the community rather than chase new technology. "It would be really easy for us to have a bunch of technical people in a room and deploy technologies that would be cool," says Jordan Davis. "But instead we said, 'Let's think about people.'" In doing so, Columbus discovered that the problems people faced were more complex and less closely related to technology than it expected. Technology provides some new opportunities, but Columbus knows it cannot provide all the answers. "Transportation technology is a very exaggerated space," Davis says. "I'm really excited to figure out what's real and what's not."

Making that distinction is the essential task faced by cities. To avoid repeating the mistakes of last century, we must take off our tech goggles and abandon unrealistic and utopian visions of AVs. It is only by recognizing a technology's limits and dangers that we can hope to attain its benefits.

3   The Democratic City: The Social Determinants of Technology's Impacts

How do democracy and politics appear through tech goggles?

This is a central question in our examination of smart cities. The previous chapter elucidated how what appear to be technical questions—How can cities improve mobility? How should cities prepare for new technology?—are actually political questions whose answers can have significant social and political impacts. But it is not yet clear whether the flaws in the visions for new transportation technology implicate other aspects of the smart city. Are the issues that emerge as cities embrace self-driving cars unique to automobiles and traffic, or do they reflect something more fundamental about a tech-centric worldview? As we will see, those under the influence of tech goggles miss not just the political nature of seemingly technical issues such as traffic but also the political nature of politics itself.

Many technologists point to direct democracy as the pinnacle of democratic civic engagement. One oft-cited model is the New England town meeting, in which community members gather to deliberate and make important decisions. In the eyes of technologists, society deviated from town meetings to our current form of government for purely practical reasons of scale and coordination: the country became too big (in both geography and population) to hold such a meeting for every decision.

"Sadly, direct democracies do not scale," explains the technologist Dominik Schiener. "Direct democracies in its purest form are simply not feasible for larger communities." Resolving this issue required numerous levels of representation, but that is a flawed solution, in Schiener's eyes. "Representative democracies scale well, but they fail to best serve citizens' best interest," he writes, explaining that rather than look out for the people they represent, politicians exist within a web of partisan political organizations and corrupt special interests. We have been stuck with the legacy of technical limitations, Schiener declares, prevented from returning to a direct democracy by "implementation barriers."[1]

Digital technology and social media appear to break those barriers and make it possible to adopt more democratic forms of governance. Complaining about the "shrill arguments" and "fights" that characterize government, former San Francisco mayor Gavin Newsom asserts: "Technology has rendered our current system of government irrelevant, so now government must turn to technology to fix itself."[2] Mark Zuckerberg promised that Facebook would "bring a more honest and transparent dialogue around government that could lead to more direct empowerment of people."[3] Napster co-founder and former Facebook president Sean Parker proclaimed that "new mediums and new media . . . will make politics more efficient."[4] Nathan Daschle (the son of former Senate majority leader Tom Daschle) promises that technology can liberate people from partisan dysfunction and disempowerment, asserting that with modern connectivity, "We can now replicate core political party functions online."[5]

Following these dreams, technologists have created apps to transform politics. Parker started the online civic engagement platform Brigade—the self-proclaimed "world's first network for voters"[6]—promising a social media–driven "revolution in political engagement."[7] Hailed as "a dead-simple way of taking political positions," the app allows users to agree or disagree on statements such as "The federal minimum wage should be raised to $15 an hour."[8] Another platform, Textizen, promises "Public engagement for the digital age"[9] by letting people respond by text message to simple questions posed by the government, such as "What's your favorite thing about Salt Lake City?"[10] "Textizen makes it possible for cities to collect and analyze data in real time," explained one observer, who added, "Textizen is reinventing the relationship between government and citizens."[11] Daschle's app, Ruck.us, created an online social network where users could discuss current events and plan political actions.

Within cities, no technology has been more heralded for transforming civic engagement than the digital services apps known as 311. Named for the phone number used in many cities to access nonemergency municipal services, these apps promise to improve civic engagement by making service delivery more personalized and efficient. Instead of needing to call city hall to ask for services, residents can photograph potholes or damaged street signs and notify their governments straight from a smartphone. Once their issue is fixed, residents receive an update from the government. As an added benefit, instead of wandering the city searching for potholes to fix, city staff can rely on their residents to notify them of issues as they arose. Dozens of municipalities (including Baltimore, Los Angeles, and Lincoln, Nebraska) have each deployed city-specific 311 apps for reporting issues within their jurisdictions.

Underlying the development of 311 apps is the belief that with the public working as the government's "eyes and ears,"[12] the efficiency of service delivery would increase and so too would trust in government. An early version of Chicago's 311 app appeared in Apple's App Store with the following description: "Engaging the citizenry in the daily administration of government will lead to more efficient allocations of tax dollars, increase transparency and increase trust in Chicago government."[13] According to IBM, "The more digital tools make it easier to interact with the government, the more confidence citizens will have in the government to provide important public services." One manager at the company proclaimed, "Social and mobile applications are fundamentally—and for the better—transforming how citizens and governments can interact."[14]

These technological developments reflect admirable goals. After all, our current representative democracy is clearly imperfect and leaves many people disempowered and disaffected. Public engagement in democratic decision making is a critical way of counteracting increasingly concentrated political power and declining trust in public institutions. Engagement is particularly important in the context of local government, where the issues at hand most directly affect daily life and where the most direct opportunities for engagement exist. In addition to allowing individuals a voice in setting priorities and policies, engagement is critical as a tool for developing their interest and capacity to become active citizens who can participate in deliberation. In this sense, it is important that citizens be able to voice their opinions to the government both as individuals and as members of civic associations—the foundations of collective action, which Alexis de Tocqueville in 1840 famously deemed the "great free schools" of democracy.[15]

Yet despite technologists' optimism and flashy new technology, they have found that "the rules of politics are not easily broken."[16] Daschle's platform Ruck.us, for instance, gave up on its mission within two years. Even Parker acknowledges that the domain of civic engagement apps is "littered with failure."[17]

New technology clearly transforms how people communicate and associate—so why hasn't it transformed democracy?

* * *

The seeds of these technologies' limitations can be found in the assumptions and priorities embedded in their design. Parker wants to make politics more "efficient," and his app Brigade is acclaimed for being "dead-simple." Schiener assesses political systems in terms of their ability to "scale." Across the United States, 311 apps draw attention for making government "easier" and "more efficient."

Technologies designed around these values fail not because they are poorly or maliciously designed, but because they do not address the fundamental challenges behind democracy and engagement. Through tech goggles we misdiagnose the core limitations on democratic decision making and civic engagement—power, politics, public motivation and capacity—as problems of inefficiency and insufficient information.

This logic suggests that politics is a coordination problem that can be solved through novel technologies, freeing us from the perils of the "dumb" city. Inspired by the rise of artificial intelligence and 3-D printing, for instance, the MIT computer scientists Christopher Fry and Henry Lieberman advocate for replacing U.S. democracy with a "Reasonocracy" that is "based on logical reasoning." In this proposed form of government, decisions are made by "Reasonocrats," representatives who "solve problems . . . using reason instead of power," as opposed to bureaucrats, "who implement government policy via fixed routine without exercising intelligent judgment."[18]

But as the political scientist Corey Robin explains, politics "is a struggle about social domination" that involves the negotiation of competing interests.[19] When half of a group wants one thing and half wants another, disagreement and disappointment are inevitable. Politicians seeking to mediate between those positions are bound to disappoint both sides. This is what Bruno Latour referred to when he wrote, "What we despise as political 'mediocrity' is simply the collection of compromises that we force politicians to make on our behalf. If we despise politics we should despise ourselves."[20] Moreover, in many areas, including education and criminal

justice, disagreements revolve around not just power or resource allocation but fundamentally competing moral visions of a good and just society.
Democracy, in other words, is not merely a project of aggregating prefer-
ences and making logical decisions.

By failing to recognize these realities of politics, technologists misdiagnose the roots of political disempowerment and dysfunction and mis-
identify the ideals that we should strive to achieve. As a result, the very
goal that technologists most desire—an efficient and conflict-free politics—
is nonsensical.

Technologists overlook numerous aspects of politics that contribute to
the issues they decry and constrain the solutions they propose. A notable
example is the design and structure of political institutions. The political
scientists Archon Fung, Hollie Russon Gilman, and Jennifer Shkabatur
note that "claims about the potential benefits of digital technologies for
democracy . . . are excessively attentive to the novel dynamics that technology enables but inattentive to the institutional dynamics of political systems." Proclamations that technology will empower the public and enable
direct democracy ignore the fact that collective action requires resources
and authority, and that "policy makers and politicians have little incen-
tive" to engage more deeply with the public. Moreover, Fung, Gilman, and
Shkabatur explain, although apps like 311 enable individuals to send information to the government, "they do not aim to create more equal, inclusive, representative, deliberative or potent forms of citizen influence over
government." The public might have a new tool to inform municipal operations, but that tool is employed within the framework of traditional relationships: the governed obtain services from the government without being
empowered with greater agency over public priorities or decision making.
In fact, it is precisely because outreach apps such as 311 and Textizen are
"compatible with existing incentives and institutional constraints"—rather
than revolutionary—that governments have eagerly adopted them.[21]

Notions of politics and democracy as coordination problems also mis-
represent what gets people engaged in politics and what barriers keep them
out of it. Engagement apps strive to simplify civic interactions as much
as possible, as if the only obstruction to civic engagement were the time
it takes to speak face-to-face with neighbors and public officials so that
lowering communication barriers would enable true democracy to emerge.
Schiener espouses this view, declaring that "making the entry barrier as low

as possible . . . will much more likely satisfy a large portion of the population and lead to an overall better governance of the country."[22]

Such assumptions fly in the face of well-established lessons from civic associations: attempts to transform democracy by simplifying engagement are doomed to failure. As the political scientist Hahrie Han explains in *How Organizations Develop Activists*, engagement efforts that are designed to be quick and easy fail to foster meaningful participation and citizenship. In a multiyear, nationwide study of a dozen civic groups, Han concluded that a reliance on "transactional mobilizing" plagued low-engagement organizations, while an emphasis on "transformational organizing" allowed organizations to develop large cadres of dedicated participants and in turn have more significant political impacts.[23]

Han describes transactional mobilization tactics as those that conceptualize participation "as a transactional exchange between the activist and association."[24] A civic group organizes opportunities to get involved in an issue—call an elected official, show up to a protest—while the individual provides time and energy. From this perspective, the best way to foster engagement is to make participation as quick and simple as possible. This task becomes increasingly easy with new digital tools that drastically lower the barriers to communication.

But when reaching many people is so easy, even high numbers of participants can be a misleading indicator of strength. Many people might fill out a survey, but be hesitant or unprepared to become more active citizens. By lowering barriers as much as possible to meet people where they already are, in other words, transactional tactics fail to empower individuals or motivate them to take further action. In Han's survey, organizations that relied on transactional mobilizing "became trapped," constantly struggling to maintain engagement.[25] And because elected officials often recognize (and discount) political action that is "cheap talk,"[26] these organizations lacked the capacity to effect change.

In contrast, the high-engagement organizations in Han's research utilized transformational organizing strategies that "cultivate people's motivation, skills, and capacities for further activism and leadership."[27] These groups also deploy mobilization tactics to get people involved, but with an emphasis on civic development rather than transactions: people who engage are recruited to take on more responsibility and taught how to lead future efforts. Through this process, high-engagement organizations teach members to reflect on the value of their work and enmesh them in social relationships that act as critical motivators to stay engaged. In contrast, participation wanes in low-engagement groups because individuals lack the sense of purpose and collective identity to stay motivated.

Engagement apps are exemplary transactional mobilizers: they minimize barriers to submitting requests and opinions in order to maximize the quantity of interactions, but they do not cultivate deeper engagement or create community among users. They are poised to fall prey to the same political impotence as Han's low-engagement groups: just as those organizations were constrained by their members' lack of motivation and civic skills, so too will digital platforms that rely on opinions or service requests become trapped by the limited nature of the participation they encourage.

In other words, reporting a pothole will not suddenly compel someone to vote—or run—for her local school board, nor will doing so grant her a more important voice in policy debates. Research in Boston has found that residents who receive timely responses to service requests are more likely to submit requests in the future, but there is no indication that this translates into other civic behaviors (e.g., voting, joining neighborhood groups) or attitudes (e.g., increased trust in government).[28] In fact, a separate analysis in Boston found that 311 reporting reflects hyperlocal personal needs rather than civic motivations: more than 80 percent of reports were in the immediate vicinity of the reporter's home. The study concluded that one "should not treat 311 reporting as a proxy for activities like voting or volunteering,"[29] and a similar study in New York reached the same conclusion.[30]

Transactional technology solutions for civic engagement also ignore the structural social and political factors that empower certain voices more than others. In *The Unheavenly Chorus*, the political scientists Kay Schlozman, Sidney Verba, and Henry Brady describe substantial and persistent "disparities in political voice across various segments of society," in which "the affluent and well educated are consistently overrepresented" both online and off.[31]

Beyond the most salient barriers that contribute to this disparity—education, skills, money, time, and internet access—are personal experiences that shape people's ability to assert themselves in public life. As Nancy Burns, Schlozman, and Verba note in *The Private Roots of Public Action*, U.S. women are still less politically active than men almost a century after first being allowed to vote. This discrepancy is due in part to inequalities associated with social attitudes and opportunities that at first glance appear unrelated to civic participation: for example, men "are more likely than women to get the kinds of jobs that develop civic skills and to gain positions of lay leadership." As a result, women "are less likely than men . . . to be politically interested, informed, or efficacious."[32]

Some segments of the population do not engage with civil society because they have been systematically excluded from it. The legal scholar Monica Bell observes that "many people in poor communities of color [have an intuition] that the law operates to exclude them from society." Underlying this is a perception of what Bell calls "legal estrangement": African Americans don't just perceive that they are unfairly treated by police and the law but also, more broadly, "often see themselves as essentially stateless—unprotected by the law and its enforcers and marginal to the project of making American society." One teenager in Bell's study reported feeling socially and politically powerless, asserting, "[Y]our voice basically doesn't matter."[33]

If poor, minority communities recognize themselves as excluded "from the protection of the law and from their rightful place in society"[34]—not to mention regularly harassed and shot by police, who are among the most visible of the "street-level bureaucrats"[35]—what difference does it make if they can contact the government using an app? In fact, a 2016 study found that "[p]olice misconduct can powerfully suppress" calls to 911 among African Americans—indicating that systemic disempowerment and abuse are critical barriers to trust in government and civic engagement.[36]

* * *

While no app could be expected to solve these myriad barriers to civic participation, that the creators of engagement apps provide no response to or reckoning with them indicates the severe limits of tech-centric perspectives: the innovators do not recognize these problems, or, if they do, they believe that technology can solve them. In this sense, the myopic focus on making civic engagement and governance more efficient results from the belief in a sort of "information fallacy": government is flawed because it does not have enough information about citizen perspectives and infrastructure conditions, in large part because barriers to entry for the public are too high. Reducing these barriers should therefore provide the government with more information, making it more efficient and, hence, effective.

The trouble with this perspective is not that information is useless, but that focusing solely on information as the way to improve urban democracy completely ignores the roles of politics and power. Democracy means far more than just allowing people an easy way to voice their needs and opinions so that services can be provided efficiently—it requires structuring society so that all people "stand in relations of equality" and can pursue "collective self-determination by means of open discussion among equals, in accordance with rules acceptable to all."[37]

By casting political problems (how to allocate resources across the population) as coordination problems (how to efficiently respond to constituent requests), tech goggles obscure and aggravate existing inequities. Meanwhile, the focus on efficiency as a solution distracts us from the ways that power structures and civic processes systematically exclude and diminish the weight of certain voices. Those distortions cannot be rectified through technology—they will be resolved only by changing laws and institutions.

Compounding this issue is that making engagement efficient, and in particular emphasizing efficient service delivery through 311 apps, diminishes the nature of citizenship by suggesting to the public that government exists to address personal needs, as if it were a customer service agency. Promising to quickly repair every pothole elides the reality that government has limited resources that often must be allocated to other issues and other people. Such promises will therefore generate citizens-as-disappointed-consumers and impair the "willingness of people to accept the collective responsibilities of citizenship," explains the political scientist Catherine Needham.[38] "The fundamental danger is that [treating citizens like consumers] may be fostering privatised and resentful citizens whose expectations of government can never be met, and cannot develop the concern for the public good that must be the foundation of democratic engagement and support for public services," she writes.[39]

Focusing on customer service may in fact "exacerbate political inequalities even as it improves some aspects of service production and delivery," writes the political scientist Jane Fountain.[40] Because customer service relies on meeting a customer's demands and expectations, social groups with less power and lower expectations—poorer "market segments," within the frame of customer service—will also receive lower-quality service.

The inequities inherent to efficiency- and customer-service-oriented engagement technology are clearly demonstrated by 311 apps—despite appearing to be value-neutral, the efficiency embedded in these apps benefits some groups more than others. The apps are designed around a predetermined set of request types that the public can submit; the most common categories involve potholes, street lights, graffiti, sidewalks, and trees.[41] The apps streamline access to municipal services for those whose needs fall within these limited categories, but they do little to help those whose needs transcend them. In Boston, for example, one study found that "Black and Hispanic respondents both reported wanting to use the system to connect with neighbors" and that these communities "were less likely to report public issues."[42] As a result, 311 reports are heavily skewed toward the wealthy, white neighborhoods whose residents have the greatest propensity to report issues rather than to those with the most need.[43]

More fundamentally, despite being hailed as empowering the public, efficiency-minded technologies provide no avenue for the public to request better schools, improved bus service, or less heavy-handed intervention by police—in other words, to voice needs that would require difficult and substantial (and therefore inefficient) political reforms. In fact, 311 apps largely prioritize the same types of issues as discriminatory "quality of life" or "broken windows" policing, leading to the criminalization of minorities in gentrifying neighborhoods[44] and (via a local landlord's 311 report about an "Eric" selling cigarettes outside his building) the killing of Eric Garner.[45]

Through these design choices, 311 apps can reduce the scope of collective experiences and frustrations that are essential to a unified body politic. If certain groups have their needs addressed quickly and easily, they are unlikely to recognize the challenges that other groups with deeper problems face. When an upper-middle-class white woman enjoys a frictionless process in requesting that a pothole in front of her house be repaired, she may further buy into the premise of the city-as-customer-service-agency—pleased that the city solved her problem and eager to request more services, but uninterested in engaging with her government or community in more meaningful ways. She will be unaware not just of the more intractable problems that exist in many poor and minority communities but also of the significant friction that those residents experience when they attempt to persuade the government to address their concerns.

Suppose instead that no 311 apps exist—that there is no easy way for someone to get that pothole in front of her house repaired. Perhaps now she must call the public works department, which informs her that there are even worse potholes in another part of town that need to be filled first. She begins to see that her issue, although experienced individually, is part of a collective problem: many people have potholes in front of their homes, because the government has insufficient funding for infrastructure repair. Frustrated by the large backlog, maybe she gathers her neighborhood into a civic group. Partnering with a local representative, they organize to place a referendum for a tax to support infrastructure repair on the next municipal ballot. Through this work, the civic group also comes into contact with those trying to address the community's more systemic issues, such as inadequate bus service and underfunded schools in the predominantly black neighborhoods, and joins forces with them to advocate for change.

This is certainly more work than submitting a 311 report to get a pothole filled and requires a higher level of engagement than many people desire. In a utopia, perhaps everyone really would need nothing more from government than efficient delivery of services. But in the real world, where there are vast inequalities in the scope of issues that different groups face and in whose demands are prioritized by public officials, it is deeply undemocratic to provide the more advantaged with opportunities to streamline their already privileged relationship with government while largely ignoring those with more substantial and intractable needs.

Moreover, positioning 311 apps as the solution for civic engagement obscures the real determinants of antidemocratic power in cities. As the social scientist Jathan Sadowski and the lawyer Frank Pasquale explain, "Time spent organizing to deploy a 'platform for citizens to engage city hall, and each other, through text, voice, social media, and other apps,' is time not spent on highlighting the role of tax resistance by the wealthy in *creating* the very shortage of personnel that smart cities are supposed to help cure by 'force multiplication' of the cities' remaining workers."[46] With more public funding for social services, in other words, there would be no need to strive with such urgency to make every public institution more efficient.

If anything, however, the logic of data and algorithms justifies and exacerbates austerity. In a 2016 editorial, the economist Rhema Vaithianathan (who has developed machine learning models that Pittsburgh uses to predict child abuse)[47] argues that "[b]y 2040 Big Data should have shrunk the public sector beyond recognition." Vaithianathan proposes that data should replace civil servants and, even better, that "[t]he information and insights will be . . . , ideally, agreed by all to be perfectly apolitical."[48] Viewing government through tech goggles, Vaithianathan assumes that municipalities do little more than the mechanical tasks of monitoring services and gathering information about those services. According to this logic, an emphasis on data can make government simultaneously more efficient and more socially optimal. Taking this position, Vaithianathan advocates for severe cuts in the public sector—framing her advocacy not as a political argument but as a purely technical and therefore "apolitical" one.

Technological solutions thus provide a way to eschew meaningful discussions about our political values and how to realize them. Just as self-driving cars obstruct conversations about improving livability and reducing congestion through better urban design and public transit, so engagement apps stymie consideration of systemic changes that would more significantly empower the public. Striving to lower barriers for engagement through technology takes for granted that there are no more worthwhile avenues to meaningfully increase civic participation. Yet this is clearly not the case: if inequalities in public and private life discourage many people from participating, then reducing these inequalities would engender more widespread and substantial engagement. Abolishing practices and institutions that systematically marginalize certain communities would do far more to increase civic trust and participation than developing an app that these groups can use to report graffiti.

But the problem is not just that better solutions are possible: through the tech goggles cycle, civic engagement apps alter our conceptions and practice of democracy. Viewing democracy through tech goggles, people overestimate the impact of technology and ignore the complex social and political factors that shape civic engagement. They see high barriers to communication—a problem that technology is more than capable of solving—but are blind to other limiting factors, such as community capacity to organize and minimal political incentives to redistribute power. Technologies are then designed so that city residents have seamless opportunities to contact government, but interactions are limited to those that make the government into a more efficient service provider. Such technology shapes the behavior of citizens and public officials in ways that entrench the beliefs that problems of governance result from poor coordination, that the point of government is to efficiently provide services, and that the fundamental political challenge of living in a city is dealing with basic service needs. As apps cast simply stating an opinion as the primary form of participation, people may see no reason to organize, build coalitions, or develop legislation. It is only through a focus on technology and efficiency that such impoverished views of democracy and politics could take hold.

Technologists are correct, of course, in asserting that their tools can reduce barriers to information and participation—anyone who uses email and social media can attest to this. In certain contexts, such as when municipalities need information about conditions across the city, citizen participation can provide valuable aid. In Detroit, residents used a mobile app to report the conditions of over 400,000 properties, providing the city and local nonprofits with precise data to guide urban revitalization efforts.[49] New York City relied on its 311 system in the aftermath of Hurricane Sandy to identify the locations of downed trees and other issues.[50]

But we should not confuse such interactions with meaningful civic engagement or public empowerment. Each relies on transactional information sharing and traditional relationships between citizens and government. Seen through tech goggles, which obscure issues of justice, civic identity, and power, such dynamics may appear to solve the problem of civic engagement. Yet a more thorough accounting of the barriers to civic engagement would echo the words of the open government activist Joshua Tauberer: "Governance is about power. Power is a social thing, not a technological thing. Websites don't magically give people power."[51]

Mitch Weiss, who helped create the first 311 app as chief of staff in Boston, has learned a similar lesson. "We know from our data that trust in government is at historic lows," he says with a grimace. "I don't think that trying to make government ever more efficient and treating our citizens like customers and trying to serve them better and better as customers is actually the answer. There are big decisions that we need to make in public life that are not going to be solved by technology, and we need citizens to engage on those decisions. And if you use technology to push them further and further away, they're not going to be able to make those big decisions with you."[52]

* * *

That's where Steve Walter comes in. Despite being a technology researcher for the City of Boston, Walter is far more fascinated by the transformative power of play than by the capabilities of new technology. He illustrates the value of play with a simple gravity experiment. He holds a pen in the air, drops it, and watches it fall. The pen lands on the floor and rattles around. "I'm learning about gravity!" Walter exclaims. "I'm playing and learning at the same time. That's what play does."[53]

Games and playfulness have long captured Walter's imagination. "We've all had those experiences where we became engrossed in a game and felt so alive," he says. "But it's typically for stupid reasons—to beat a stupid game! Just imagine if you could feel that way toward helping another person."

With a background in media and user design, Walter provides an important perspective in city halls: intense focus on the lived realities of people who live and work in the city. "The urban environment is an experience— it's not just being in a dense area," he explains. "It's about how you interact with other people, and how you do things with others. We have to take into account human experience." In this setting, Walter says, much of urban life's value comes from the ability to play—to derive new meaning by exploring, questioning, and pushing the bounds of experience.

Play has long been considered essential to a healthy democracy. The philosopher Marshall Berman declares, "Any society that takes the rights of man and citizen seriously has a responsibility to provide spaces where these rights can be expressed, tested, dramatized, played off against each other."[54] Berman points to the music video for Cyndi Lauper's 1983 hit song "Girls Just Want to Have Fun"[55] as emblematic of these values. The video portrays Lauper and her friends singing and dancing through the streets of New York City. Along the way, they pick up onlookers and companions (ranging from a black construction worker to a stuffy white businessman) to create a joyous dance party that mirrors the even more resplendent and city-unifying parade from the 1986 film *Ferris Bueller's Day Off*.[56] Through song and dance as a form of play, Lauper and Bueller are both, in Berman's words, "transforming the life of the street itself, using its structural openness to break down barriers of race and class and age and sex, to bring radically different kinds of people together."[57]

But as cities rush to adopt technology and make civic engagement efficient, Walter sees play being stamped out in favor of increasing efficiency. While many in his field hail new technology as encouraging civic engagement, Walter decries it "as a novelty to increase participation" without making that participation more meaningful and empowering.

Combining the spirit of play with the possibilities of digital technology, Walter wants to "use games to create new forms of civic action." His goal is not to design a game that can be won but to design a process that generates reflection, empathy, and learning. In a neighborhood planning process, for example, Walter believes that "game mechanics can create a more empathetic understanding of the other stakeholders." He explains, "I as a young white man might play as the eighty-year-old Asian immigrant who has a different set of needs than I do." Such exploration can lead to newfound understandings and create opportunities for dialogue across groups. Walter believes that it can also make engagement more fulfilling and effective. "If we can make the process intrinsically enjoyable, we will get better outputs. If people want to be a part of it, they will put more effort into it."

Several years ago, Walter joined forces with a kindred spirit: Eric Gordon, a civic media professor at Emerson College in downtown Boston and founder of its Engagement Lab. As Gordon observed the increasing use of technology as a means of civic engagement, he grew concerned that "engagement is too often conceived as simply making available opportunities for official transactions . . . rather than enabling citizen-to-citizen connections or meaningful feedback."[58] As Gordon saw it, the technological focus on making civic engagement efficient creates a "systematic blindness to the responsibility of government to cultivate dialogue, meaning and dissent."[59] He provides an example: deliberation is a deeply inefficient process, but one that is vital to a healthy and representative democracy. "How do we build that into systems where efficiency is the primary value?"[60]

Setting out to answer this question, Gordon and Walter created Community PlanIt (CPI)—an online, multiplayer game that facilitates engagement, deliberation, and decision making within communities. The game is organized around a series of weeklong missions, each of which focuses on a particular issue and consists of challenges such as trivia questions, problem solving, and creative exercises. In completing these activities, participants are prompted to consider the views and perspectives of others, all while attempting to accumulate points and influence within the game. The goal is not to push users toward a particular outcome but to provide an environment where the community can come together and deliberate.

Clearly, Community PlanIt is not the typical civic engagement technology platform. But it has been successfully deployed in numerous contexts, including a 2011 visioning process for the Boston Public Schools and a 2012 master planning process in Detroit. Both cities saw remarkable results that far surpass the benefits provided by typical engagement apps. An evaluation of CPI in these two cities found that the game "creates and strengthens trust among individuals and local community groups" and "encourages interactive practices of engagement."[61] In Detroit, where numerous engagement tactics were deployed as part of the long-term planning effort, Community PlanIt was voted the one that made participants feel most hopeful about the future.[62]

Instead of being gamified with a rigid structure that funnels users to predetermined ends, CPI embraces play to enable exploration and deliberation. Every user is tasked with responding to open-ended prompts, and in order to see the responses of others, one must first submit one's own answer. Such game mechanics lead to positive and reflective deliberation that one participant called "the back and forth that you don't get in a town hall meeting." Players also noted that the game encouraged them to reflect on their own opinions and appreciate alternative viewpoints. "I think it forced you to really think about what you wanted to say in order to see other people's opinions," said one participant. "Whenever I found out that I was like the minority . . . it just made me think of why do people think the other idea is better," added another. "I put my comment and someone disagreed with it," remarked another player, before adding, "I don't really know who's right, but I feel like it made me really think about what I thought prior."[63] Through these interactions, players developed their capacities to reflect on their positions and emerged with deeper trust in the community.

What makes CPI most special in Walter's eyes are the "mechanics within the game that allow people to appropriate it." The conversations that emerge from the game are often not those that public officials have in mind at its start. For instance, in a small Massachusetts city where a local planning commission came in with a set of survey questions, the public turned the conversation to an entirely unexpected topic: trash pickup. Even though this was not part of the game's prescribed context, Walter explains, "it was just the thing that kept emerging over and over again. They wanted to talk about that."

"This is the best part of a system that uses game mechanics but ultimately cares more about play than about structure," Walter says. "You provide some structure but you allow people to push it in the direction that they want it to go."

Observing the impacts of Community PlanIt led Gordon and Walter to develop an appreciation for what they call "meaningful inefficiencies." In contrast to "mere inefficiencies" that cause systems to lag unnecessarily (there is little value in inefficient snow plowing, for instance), they explain that "inefficiency becomes meaningful" when it enables "citizens [to] share in a give and take of experience and increase their range and perception of meanings with each other." Meaningful inefficiencies make possible "civic systems that are open to the affordances of play . . . , where users have the option to *play* within and with rules, not simply to *play out* prescribed tasks."[64]

Tech goggles, of course, fail to differentiate between mere and meaningful inefficiencies: all inefficiency is bad. "When the application of technology to civic life is celebrated purely for its expediency, transactionality, and instrumentality," Gordon and Walter lament, civic actions such as deliberation, dissent, and community building "are potentially sidelined." And "by tacitly managing the possible forms of self-government and fields of action available to citizens," they argue, governments retain their control through technology "far more efficiently and pervasively [than possible] through external force."[65]

This is a far cry from the bottom-up revolution that many hoped the internet and social media would enable, and Gordon says that change can come only from "a culture shift. I think it's really important that it becomes more prominent to think about the institution of cities as not simply being about providing service, but being about creating livable contexts," he says. "That is not simply a matter of infrastructure and service delivery—it's a matter of creating cultures where dialogue and deliberation matter. That isn't often generated simply through transactions."[66]

As Community PlanIt demonstrates, meaningful inefficiencies need not be eradicated by technology. Systems like Community PlanIt will not solve every community challenge or provide citizens with greater influence over public decisions, but they are far more capable of promoting civic engagement than the typical apps we have seen. Rather than funneling citizens toward simple and transactional behaviors, meaningful inefficiencies allow for play that transforms civic perceptions, motivations, and capacities.

In this sense, meaningful inefficiencies can be seen as kindling that fuels transformational organizing.

For when it comes to developing technology to create more democratic cities, embracing values and policies other than efficiency—that in fact are often rooted in meaningful inefficiencies—is paramount. For as Gordon and Walter explain, the fundamental question is not "how can we make civic life more efficient with technology" but rather "how can we use technology to make civic life more meaningful."[67]

* * *

Technology, as we have seen, does not exist in a vacuum: despite the hopes of many, it does not demolish institutional and political structures on its own. Instead, technology can support deliberation and capacity building only to the extent that its users are actually granted meaningful voice in developing public priorities and policies. The challenge for cities attempting to improve engagement is not to deploy cool new technology but to create civic spaces that empower the public. Power shifts not when a new technology makes existing processes and interactions more efficient but when those processes and interactions are restructured to give the community greater influence over local governance, whether that is achieved through technology or not.

One initiative that provides this voice is participatory budgeting (PB), a process in which the government empowers residents to directly determine how a portion of their municipal budget will be spent. Through PB, residents work with one another and government officials to develop proposals for projects that the government could fund. After projects are selected through a democratic vote, the government works with the community to implement them.

Compared to the traditional method of allocating budgets, in which the public has little or no influence over how public money is spent, participatory budgeting represents a stark shift in power. By "creating a new process for how citizens and institutions share information, interact, and make public decisions," writes the political scientist Hollie Russon Gilman in *Democracy Reinvented*, "participatory budgeting has the potential to strengthen local democratic practice and to alter the current relationship between citizens and local government."[68]

In addition to allowing residents to collectively develop and allocate money to projects that address community needs, PB creates a rare environment of deliberation and knowledge transfer. As members of the public work with government officials and develop municipal projects, they gain a more thorough understanding of government's role and limitations. Through this process, citizens create social capital, develop a sense of leadership and agency within their community, and come to appreciate the multiplicity of needs and values that public policy must balance. Although occasionally frustrating, these aspects of PB are essential to its appeal: one participant proclaimed PB "the most fulfilling mode of civic engagement I have ever been part of."[69]

None of this means that participatory budgeting is commonplace or easy to implement, however. PB has been used in the United States only since 2009 (though the practice's roots trace back to Brazil in the 1980s), and the process is incredibly time-consuming. To develop and select projects, people must attend numerous meetings, which many are unable or unwilling to do. Those with inflexible job schedules and childcare responsibilities are particularly limited by these time requirements. For PB to encompass larger budgets and engage more people, notes Gilman, "it will need to become less resource-intensive."[70]

Digital tools for information sharing and communication could make PB vastly less burdensome by streamlining deliberation, but Gilman warns that reforming the process with a focus on efficiency could "weaken the commitment to face-to-face engagement that makes deliberative exercises worthwhile." Gilman found that "citizens primarily sustain their involvement in [PB] because of [its] civic, not material rewards," which "are much more difficult to acquire online." For example, working with public officials is "among the most rewarding aspects of the process" but would be almost impossible to replicate within a technological platform that is designed to make conversations efficient. Similarly, replacing group deliberations with electronic communication is likely to shrink the range of topics and participants while also reducing the community building that participation creates. In these ways and more, asserts Gilman, technology may dilute PB by preventing participants from "experiencing the painstaking rewards and gaining the kinds of knowledge that come from in-person participation in civic dialogue."[71]

Gilman is right to fear that adopting typical civic engagement technologies—replacing in-person deliberation with an online platform for efficiently making decisions, for example—would diminish PB's value by shifting the emphasis from transformation to transaction. After all, Hahrie Han's research highlights the value of transformational organizing that develops people's capacities as activists and leaders. Transactional mobilizing can aid these efforts by bringing people to the table, but it cannot on its own generate sustained civic engagement. The same holds for participatory budgeting: beyond providing new mechanisms for the public to influence municipal decisions, PB draws its value largely from the transformational processes it involves. Participants emerge having learned how government operates and having developed their capacity to effect change. Even when the process was frustrating, Gilman writes, participants remained committed because they "were forging . . . a collective identity that sustained their involvement."[72]

But perhaps Gilman assumes that technology is anathema to PB only because civic technology is typically developed with such a laser focus on efficiency. Notably, Han does not dismiss transactional approaches entirely. Although they are insufficient on their own to sustain efficacious organizations with highly engaged members, such approaches are nonetheless a vital tool for drawing people in. Transactional technologies that make involvement in PB easier could be effective tools to engage a larger portion of the public. In fact, evidence from Brazil suggests that certain efficiency-minded approaches such as online voting and text alerts for upcoming events can broaden participation in PB's more transactional components and "can be seen as the gateway for politically inactive or less active citizens."[73]

Meanwhile, Gordon and Walter's notion of meaningful inefficiencies explains what Gilman regards as "the paradox presented by an innovation [participatory budgeting] that actually both creates and depends on what some might consider inefficiency."[74] When Gilman attributes PB's value to its being inefficient compared to conventional budgeting, she is picking up on the role of meaningful inefficiencies as a source of civic rewards. From this perspective, there is no paradox at all: it is precisely because participatory budgeting requires taking a more deliberative route to the end result that it is so transformative. Thus, while Gilman is correct that most civic technologies would be detrimental to PB, Community PlanIt demonstrates that technology could aid PB's more deliberative components without warping them if it were developed and deployed to support rather than extinguish meaningful inefficiencies.

Recognizing its need to make PB more accessible, the small California city of Vallejo has begun adopting exactly these types of technologies. In 2015, Vallejo teamed up with Stanford's Crowdsourced Democracy Team to deploy a platform that facilitates online voting. In 2017, Vallejo also partnered with the Social Apps Lab at the University of California, Berkeley to incorporate into PB an additional platform called AppCivist that provides a centralized place to develop proposals, keep track of project updates, and communicate with city staff.

Vallejo has implemented these new conveniences to bolster rather than eliminate core components of participatory budgeting. As Vallejo's PB project manager Alyssa Lane emphasizes, "technology cannot replace face-to-face interaction." She notes that even though project teams could develop proposals and city staff could provide feedback through AppCivist, most preferred to do so in person, in more personal and thorough conversations. Project teams instead used the platform primarily to document progress between in-person meetings and to keep people who had to miss a meeting informed. "I think face-to-face helps them iron things out," Lane says. "I don't see that going away anytime soon."[75]

Of course, with or without technology, participatory budgeting is not a panacea for the lack of urban civic engagement and democratic decision making. Its time-consuming nature significantly limits who can participate and how widespread the practice can become. And even for those who do participate, deliberation is not foolproof: though it clearly can promote civic rewards, it does not preclude antidemocratic outcomes.[76] Moreover, participatory budgeting has thus far been restricted to relatively minor allocations of money to relatively apolitical projects. If PB is to "reinvigorate local democracy," asserts Gilman, the process "must encompass major budgetary questions, up to and including urban redevelopment, zoning, and social welfare spending."[77] Such a change will subject PB to a far more intense level of scrutiny, which may make it hard for PB to retain its essential character. Finally, many important municipal decisions are made outside the realm of budgeting, whether in implementing approved projects or in developing priorities and legislation. Citizen empowerment efforts that neglect these nonbudgetary components of politics and power will be rendered impotent, as important decisions can be shifted so that they escape PB's purview.

Nonetheless, thoughtfully integrating technology into participatory budgeting represents precisely the form of innovation that embodies the Smart Enough City. In improving a valuable but burdensome process, cities need to judiciously distinguish between mere inefficiencies that technology can and should overcome—creating opportunities for transactional mobilizing that engages more people—and meaningful inefficiencies that allow transformational organizing to occur—inefficiencies that technology should not be deployed to mitigate. That is the key distinction which technologists far too often miss: while all inefficiencies make democratic governance more challenging, only mere inefficiencies should be eradicated. That meaningful inefficiencies represent a challenging but necessary component of democracy is a paradox beyond the comprehension of technophiles.

While technology should not eliminate the necessary difficulties of in-person deliberation, it can enhance deliberations with improved information sharing, help increase the number and diversity of participants, and mobilize the community through electronic voting. Unlike 311 and other engagement apps, which promote transactional mobilizing in a vacuum, PB technologies of the sort used in Vallejo are enmeshed in a transformational experience and are designed to move people up the ladder of participation. This combination of eliminating mere inefficiencies while fostering meaningful inefficiencies will trim the fat from participatory budgeting, making it more sustainable without hindering its ability to empower the public and cultivate civic identities.

In Vallejo, for instance, participatory budgeting is being used to rebuild the trust in government that cratered after the city went bankrupt in 2008. "We have a few projects that probably would not have been funded by the city in a normal budget process," says Alyssa Lane. "But even more so than projects, I think I've been the most excited by watching specific delegates and committee members blossom." Lane shares the story of a man who began showing up to PB meetings who had never before been actively engaged in the community. Initially very quiet and introverted, by the end of the process this man had developed a project that was funded through PB, and he was in regular contact with public officials. He remained engaged in future PB cycles as a volunteer. "It was very inspiring to see the evolution of his civic engagement," Lane says. "There are many stories like that."

* * *

Returning to the question that began this chapter, we now see that technologists hold the deeply flawed and dangerous conception of democracy as predominantly a technical problem. More importantly, it is now clear that the issues explored in the previous chapter reflect not something specific to how technologists think about cars but rather their core misunderstanding of technology's value for society. Like self-driving cars, civic engagement apps are not inherently bad—but the belief that technology can solve long-standing social and political dilemmas entrenches existing structures and inequities under a shiny facade of innovation.

Participatory budgeting demonstrates that the most important innovations come in the form of programs and policies that alter social conditions and relations rather than in the form of new technology, which typically buttresses existing structures and relationships. For although PB will benefit from the thoughtful implementation of technology, its real innovation lies in providing the public with a new source of power and a deliberative space in which to wield it. Participatory budgeting thus demonstrates to Gilman that "innovation can come in many different forms, including in the makeup of the participants and in the process structure." Such an initiative would have no place within the smart city, which equates innovation with technology and therefore is primarily concerned with how new technology can make engagement more personalized and efficient. It is precisely because PB does not follow this standard approach of chasing technology-cum-efficiency that it represents for Gilman "an unlikely exemplar of twenty-first-century innovation."[78]

Instead of lowering the barriers to the simplest forms of civic engagement, the Smart Enough City reforms civic processes first and then deploys technology to improve their implementation. For technology on its own will not make cities meaningfully more democratic—it must be deployed to advance programs and policies that empower the public. This is akin to the previous chapter's conclusion that having urban development goals in place is a prerequisite to taking advantage of new transportation technologies. In the chapters to come, we will continue to explore how nontechnological innovation and long-term planning lay the foundation for effective use of technology and drive the emergence of Smart Enough Cities.

We have also begun to see how the social and political impacts of technology depend largely on the values embedded in its design and functionality. In the previous chapter the fundamental challenge that cities faced was to prepare for and adopt new technology, but here we encountered the even murkier and more consequential realm of how cities should develop and deploy new technology. Of course, determining what values to prioritize when creating technology, just like developing the policies and structures that the technology supports, is a matter of politics. We cannot escape these thorny issues simply by using technology. That will be a central theme in the next chapter.

4 The Just City: Machine Learning's Social and Political Foundations

In 2012, while reviewing reports of recent crimes, a data analyst for the Cambridge, Massachusetts, Police Department (CPD) noticed a striking pattern of thefts: laptops and purses were repeatedly being stolen on Tuesday and Thursday afternoons at a local café. While no incident on its own would have indicated much, the full set presented a clear case of a thief acting systematically. Having determined this pattern of behavior, the analyst could predict when and where the thief would strike next—and catch them in the act.

"We provided the detectives with Tuesday afternoon between four to six as the best timeframe," recalls Lieutenant Dan Wagner, the commanding officer of CPD's Crime Analysis Unit. "The detectives sent a decoy—their intern—with a backpack and a computer hanging out of it. They were there a short while, and they see the guy steal the laptop and make an arrest."[1]

It sounds straightforward, but such patterns typically go undetected as analysts struggle to find patterns within large databases of crimes. Indeed, it took several weeks for CPD to identify the café crime series, and the identification was possible only because the analyst happened to remember having seen records of similar crimes when new thefts were reported. Despite being glad that the café thief was stopped, Wagner realized that this ad hoc, individualized approach was quite limited. After all, he explains, "No crime analyst can truly memorize and recall a full historical database of crimes."

CPD's Crime Analysis Unit was founded in 1978 as one of the nation's first such teams by Wagner's mentor Rich Sevieri, who has overseen the unit's transformation from analyzing crime using pin maps and punch cards to databases and predictive models. Sevieri began his career as a journalist, and despite multiple decades enmeshed in data he still focuses on "the five Ws": who, what, when, where, and why. Even as policing becomes increasingly focused on data and algorithms, Sevieri's analytical approach remains the same: "You have to know the motivation and you have to know the scenario around the crime."

Recognizing the futility of relying on an analyst performing manual database queries to find crime patterns, Sevieri and Wagner approached Cynthia Rudin, a professor of statistics at MIT, hoping for a way to automatically analyze crime trends. Although incidents within a crime series are known to follow patterns based on the offender's modus operandi (MO), identifying these patterns is difficult for a human or computer working alone. Crime analysts intuitively sense what characteristics might indicate that a string of crimes are connected, but they cannot manually review data of every past crime to find patterns. Conversely, computers are proficient at parsing large sets of data, but they may not recognize the subtle connections that indicate a crime series.

Sevieri believed that if they could teach a machine "the process of how an old-time analyst worked," an algorithm could detect patterns of crimes that would otherwise take the police weeks or months to catch, if they could be detected at all. It would enable the Cambridge Police to more quickly identify crime patterns and stop the offenders.

When Rudin learned how CPD had stopped the series of café thefts, she recognized that "finding this pattern was like finding a needle in a haystack."[2] An expert in designing computational systems to assist human decision making, Rudin was eager to help CPD sort through more haystacks. She and her doctoral student Tong Wang began working with Wagner and Sevieri to develop an algorithm that could detect crime series patterns among residential burglaries (an offense notoriously difficult to solve[3]).

At the heart of the algorithm was a focus on identifying the MO of offenders. For a crime like burglary, perpetrators typically exhibit a particular pattern of behavior that carries across multiple incidents. When presented with the right data, Wagner and Sevieri say that they tend to see such patterns "automatically." But they may not know where to look, and can process only a limited amount of data. What makes identifying these patterns difficult for a computer, on the other hand, is that every offender's MO is unique. Some people force open front doors of apartments on weekday mornings; others break in through a window and ransack houses on Saturday nights. Thus, instead of just teaching a computer to search for a particular pattern, the algorithm had to "capture the intuition of a crime analyst" and re-create it on a larger scale.[4]

With significant input from Wagner and Sevieri, Rudin and Wang developed a model that analyzes crime series in two complementary stages. First, the model learns "pattern-general" similarities, representing the broad types of patterns that are typical of crime series (for example, proximity in space and time). Second, the model uses this knowledge to identify "pattern-specific" similarities—in other words, the MO of a particular crime series.[5] In this two-pronged approach, the model first learns the intuitions that a human analyst would follow and then applies them to a large database of housebreaks to detect crime series.

The last ingredient that the algorithm needed was a corpus of historical data from which to learn what a typical crime series actually looks like. Because of Sevieri's stewardship, the Cambridge Police Department has one of the country's most extensive databases of crime records, with detailed information about crimes over the past several decades. This data enabled the MIT model to learn from 7,000 housebreaks that occurred in Cambridge over a fifteen-year period, with details about each burglary such as the geographic location, day of the week, means of entry, and whether the house was ransacked. The algorithm also drew from fifty-one analyst-identified crime series during this period, gaining insights about what makes a crime series stand out.

Once it had learned from the data, the model quickly demonstrated its potential to help police investigate crime. In 2012, there were 420 residential housebreaks in Cambridge. The first time the team ran the algorithm, the model identified a past crime series that had taken the Crime Analysis Unit more than six months to detect.

In a retrospective analysis, the algorithm further demonstrated its ability to inform police investigations and draw inferences that would not have occurred to a human analyst. The Cambridge police had previously identified two crime series that took place between November 2006 and March 2007. When the algorithm analyzed past crimes, however, it determined that these supposedly separate crime series were in fact connected. Despite a month-long gap and a shift several blocks north in the middle of this crime series—leading the Cambridge police to suspect that the two sets of events were separate—the MOs of the perpetrators were otherwise similar: almost every incident involved forcible entry through the front door during the workday. When Wagner and Sevieri were presented with the algorithm's assertion that these two sets of burglaries were actually connected as one crime series, they recognized that it was right. The time lag that occurred halfway through the series could be explained by the deterrence effect of more people being home during the winter holidays; the geographic shift was a response by the perpetrators to having been observed while carrying out a prior burglary.[6] Had CPD possessed this information at the time, the police could have identified and addressed the emerging crime series before it expanded further. "If you don't stop that series early," reflects Sevieri, "that's what happens."

* * *

The Cambridge algorithm draws on a set of techniques known as machine learning. These methods of predictive analytics are powerful because they can mine large datasets and examine complex trends, identifying patterns that a human investigator would struggle to uncover. As the amount of data generated and stored grows exponentially, the ability to make informed decisions using this data becomes increasingly valuable.

Consider the way that Gmail monitors your incoming emails to detect spam. Every time you receive an email, Gmail evaluates the message to determine whether it is legitimate (and should be sent to your inbox) or spam (and should be sent to your spam folder). While an engineer could predetermine specific rules that characterize spam, such as the presence of the phrase "Limited time offer" and at least two spelling errors, machine learning algorithms can analyze emails from the past to detect more subtle and complex patterns that indicate whether an email is spam.

Typical machine learning algorithms rely on "training data" composed of historical samples that have already been classified into categories. For a spam filter, the training data is a corpus of emails that have previously been labeled by people as "spam" or "not spam." The next step for Gmail's engineers is to define each email's attributes, known as features, that the algorithm should consider when evaluating whether an email is spam. Relevant features in this case may be the email's words, the address from which the email was sent (e.g., is it in the recipient's contact lists?), and the type of punctuation used. Gmail then uses a machine learning algorithm to characterize the relationship between the features and the two labels. Through a process of mathematical optimization known as "fitting," the algorithm determines how strongly each feature corresponds to spam messages. It thereby generates a formula, known as a model, that can classify new examples. Every time you receive an email, Gmail applies what it has learned. The model evaluates the email to determine whether it more closely resembles the spam or the not-spam examples from the training data, and in turn gauges how likely the email is to be legitimate.

Spam filters are just the tip of the machine learning iceberg, of course. Machine learning algorithms are behind the software that drives cars, beats world champions at games like chess and poker, and recognizes faces.

The abilities of machine learning to make sense of complex patterns and predict hitherto inscrutable events suggest to many that we can rely on data and algorithms to solve almost any problem. Such thinking led the entrepreneur and *Wired* editor Chris Anderson to proclaim in 2008 that Big Data represents "the end of theory."[7] Who needs to understand a phenomenon when there is enough data to predict what will happen anyway?

But as we saw in Cambridge, this claim could not be further from the truth. Sevieri talks about striving to "capture the intuition of a crime analyst" because the algorithm must understand the theories that a human analyst would use to inspect crime patterns. Of course, the algorithm need not follow the exact same thought process: much of machine learning's power comes from its ability to interpret large datasets using different methods than those used by people. But unless the model is provided with a basic framework for how to operate—such as what information to consider and what the goals are—it will flounder. MIT's model succeeded precisely because Rudin and Wang relied on Wagner and Sevieri's expertise to decipher the art of how a crime analyst thinks. The algorithm incorporated assumptions about what information is relevant (the day of the week, but not the temperature) and how to interpret that information (two incidents that occur on Tuesday and Wednesday are more likely to be connected than two incidents that occur on a Friday and a Saturday).

Although it may appear that data-driven algorithms do not rely on theories or assumptions about the world, in reality algorithms always reflect the beliefs, priorities, and design choices of their creators. This is true even for a spam filter. The process starts when Gmail's engineers select training data from which the algorithm can learn. To ensure that the algorithm learns rules that apply accurately to every type of email correspondence it will see, the emails in the training data must be accurately labeled and representative of the emails that the spam filter will evaluate in the future. If Gmail selects training data in which spam emails are overrepresented, its spam filter will overestimate the likelihood that an email is spam. Furthermore, selecting features for the algorithm requires some intuition regarding what attributes of an email are likely to distinguish spam. If Gmail's engineers know about one indicator of spam emails but not another, they might generate features that will capture spam emails like those they have seen but not other types.

Finally, Gmail must determine the goal that its spam filter will be optimized to achieve. Should it strive to capture every spam email, or are some types of emails more important to get right than others? If Gmail's engineers decide that phishing emails (which attempt to trick recipients into providing sensitive information such as passwords) are the worst type of spam, it can optimize its filter to catch them—but then the filter might be less able to identify other types of spam, such as emails about payday loans. As part of this calculation, Gmail must consider the trade-offs between false positives (marking a legitimate email as spam) and false negatives (allowing a spam email into your inbox). Focus too heavily on avoiding false positives, and Gmail inboxes will become cluttered with spam; focus too heavily on avoiding false negatives, on the other hand, and Gmail may wrongly filter out important messages. This decision can make or break a model: the March 2018 collision in which a self-driving Uber car struck and killed a woman in Arizona was the result of software that was tuned to diminish the importance of false positives (to avoid overreacting to obstacles such as plastic bags).[8]

When we are not thoughtful about these design choices, we risk unleashing algorithms that make inaccurate or unfair decisions. For although most people talk about machine learning's ability to predict the future, what it really does is predict the past. Gmail detects spam effectively because it knows what previous spam looked like (that is the value of training data) and assumes that today's spam looks the same. The core assumption embedded in machine learning models is that the characteristics associated with certain outcomes in the past will lead to the same outcomes in the future.

The trouble with predicting the past is that the past can be unsavory. Data reflects the social contexts in which it was generated. A national history rife with systemic discrimination has generated data that reflects these biases: when employers prefer white job applicants to similarly qualified

African American ones or men over similarly qualified women,[9] and when women and African Americans have been excluded from economic and educational opportunities, the resulting data about society, taken on its face, can seem to suggest there is something fundamental about being white or male that makes a person more qualified, educated, and prosperous. In other words, uncritically relying on data drawn from an unjust society will identify the products of discrimination as neutral facts about people's inherent characteristics.

When it comes to spam, this may not matter much—receiving junk mail is rarely more than a nuisance. But when it comes to algorithms that make more important decisions, the biases contained within training data can matter a great deal.

In the 1970s, St. George's Hospital Medical School in London developed a computer program to help it weed out applicants. Given an applicant pool of around 2,000 for only 150 spots, a program that could cut down the work of selecting students to interview had obvious appeal. Throughout much of the 1980s, this program conducted the school's initial review, filtering which students St. George's should interview. In 1988, however, the U.K. Commission for Racial Equality investigated this algorithm's use and found it to be biased: by following the computer program, St. George's had unfairly rejected hundreds of women and minorities with sufficient academic credentials to merit an interview.[10]

The algorithm did not learn this bias on its own: throughout the history of St. George's, the admissions staff had been making biased admissions decisions based on race and gender. When the admissions algorithm drew on its training data of previous admissions decisions, it inferred that St. George's considered women and minorities to be less worthy. Instead of learning to identify the most academically qualified candidates, in other words, the algorithm learned to identify the applicants that looked the most like those the school had admitted in the past. In fact, the algorithm achieved 90 percent correlation with a human selection panel at the hospital—which is why St. George's believed in the first place that the selection algorithm would be useful.

To this day, many have repeated the same mistake, relying on machine learning to make important decisions, only to realize that the models were making biased predictions. In 2014, for example, Amazon began developing machine learning algorithms to help it decide which job applicants to hire. Just a year later, the company abandoned the project when it discovered that the model was unfairly favoring male candidates.[11]

* * *

The Cambridge Police Department was not alone in believing that machine learning could improve police operations. With machine learning having attained an almost mythic status for being able to solve any problem, says the technology policy lawyer David Robinson, "people are primed for the idea that computers can cause significant improvements to whatever they're added to." It was only natural, he adds, that police departments across the country would wonder, "Why can't we sprinkle some of that magic over the difficult problem of community safety in cities?"[12]

Companies began showering the market with tools to perform "predictive policing." One of the most widely used is PredPol: software that, on the basis of historical crime records, analyzes how crime spreads between places and then forecasts that spatial process into the future to predict where the next crimes will occur. The company translates these predictions for police via an interactive map overlaid with red squares (covering 500 feet by 500 feet) at the predicted high-crime locations. If police spend time in those regions, the company posits, then they will be more effective at preventing crime and catching criminals.

"Vendors are happy to provide the impression that their systems will leverage technology to make things better," explains Robinson. PredPol has aggressively shared case studies asserting the effectiveness of its software, citing "a proven track record of crime reduction in communities that have deployed PredPol."[13] As explained by Andrew Ferguson, a legal scholar and the author of *The Rise of Big Data Policing*, predictive policing is alluring to police departments because it provides "'an answer' that seems to be removed from the hot button tensions of race and the racial tension arising from all too human policing techniques." He adds, "A black-box futuristic answer is a lot easier than trying to address generations of economic and social neglect, gang violence, and a large-scale underfunding of the educational system."[14]

Thus, in the wake of growing outrage about discriminatory police practices—including numerous high-profile police killings of African Americans—and burgeoning support for systemic police reforms, predictive policing was hailed as "a brilliantly smart idea" that could "stop crime before it starts" through objective, scientific assessments. In an interview, a former police analyst who served for several years as a PredPol lobbyist declared, "It kind of sounds like science fiction, but it's more like science fact."[15]

By now, however, when we observe such faith in technology—such a clear example of tech goggles at work—we should be skeptical and raise several questions: Can this technology actually achieve its stated purpose? What values are embedded in the technological solution? What do we overlook by assuming that this issue is a technical one?

Thorough evaluations of predictive policing tools suggest that they promise far more than they can deliver. A 2016 study led by Robinson "found little evidence that today's systems live up to their claims." His report instead asserts, "Predictive policing is a marketing term."[16] In fact, many of the statistics touted by PredPol are cherry-picked numbers that take advantage of normal fluctuations in crime to suggest that PredPol generated significant reductions.[17] As one statistician notes, this type of analysis "means nothing."[18]

John Hollywood, a researcher at the RAND Corporation (a policy think tank) who has assessed numerous predictive policing tools, calls any benefits of predictive policing "incremental at best" and says that to predict specific crimes "we would need to improve the precision of our predictions by a factor of 1000."[19] Hollywood's analysis of a predictive policing effort in Louisiana—one of the only independent analyses of predictive policing that has been conducted—found that the program had "no statistically significant impact" on crime.[20]

Despite their interest in using machine learning to prevent crime, even Wagner and Sevieri of the Cambridge Police Department are critical of PredPol. "It was the right product at the right time," says Sevieri. "Police departments were looking for a quick fix." Wagner's primary critique is that PredPol relies on an "oversimplified" model that "doesn't take into account the patterns" of crime. For example, PredPol assumes that the likelihood of crime in a location spikes immediately following the most recent nearby crime and then gradually decreases. Following a housebreak on Wednesday afternoon, PredPol predicts a higher likelihood of further crime on Wednesday night than on Thursday afternoon. But in reality, Wagner says, "a lot of crimes are serial crimes. There are patterns." A trend of weekday afternoon housebreaks suggests that the next housebreak will occur on a following weekday afternoon, not at midnight on the same day.

A more pointed concern regarding predictive policing models is whether they make racially biased predictions about where crimes are likely to occur. Supporters of predictive policing assert that the software must be fair, because it relies on data and algorithms. According to Brett Goldstein, Chicago's former chief data officer, an early predictive policing effort in Chicago "had absolutely nothing to do with race," because the predictions were based on "multi-variable equations."[21] Los Angeles Police Commander Sean Malinowski called PredPol "objective" because it relies on data.[22] Similarly, the director of Hitachi's crime-mapping software declared that the program "doesn't look at race. It just looks at facts."[23]

But the "facts" of the matter—in this case, crime statistics—are well known to be "poor measures of true levels of crime," writes the criminologist Carl Klockars. Because "police exercise an extraordinary degree of discretion in deciding what to report as crimes," Klockars explains, police statistics "are reflective of the level of police agency resources dedicated to [the] detection" of particular types of crime, rather than the actual levels of crime across society.[24] In other words, what appear to be facts about crime are largely facts about police activity and priorities.

For years, police have disproportionately targeted urban minority communities for surveillance and arrests, leading to decades of crime data that reflect this discriminatory treatment.[25] Police predominantly patrol black neighborhoods and possess significant discretion regarding when and why to arrest someone.[26] Many incidents that police never observe, act on, or even target in white communities are recorded as crimes in black neighborhoods.[27]

This is what makes The New Inquiry's "White Collar Crime Early Warning System" such a wonderful piece of satire. The magazine developed a model, using similar technical approaches as predictive policing tools, that predicts where financial crimes are likely to occur.[28] In Chicago, for example, whereas most crime maps show hot spots in the predominantly black and brown south and west sides, the hot spots for white-collar crime are in the central business district ("The Loop") and the primarily white north side. That these maps—and in fact the very idea of using algorithms to proactively target financial crimes—are so striking brings to light an oft-overlooked aspect of the criminal justice system and machine learning–based reform efforts: our very selection of the crimes that ought to be aggressively monitored and enforced rests in part on racist and classist notions of social order.[29]

Thus, even if a machine learning algorithm is not hard-coded to exhibit racial bias, the data from which it learns reflects social and institutional biases. As we saw in chapter 3, differential rates of reporting on 311 apps could lead one to conclude that all of a city's potholes are in the white, wealthy neighborhoods. And as we learned from the admissions algorithm at St. George's Hospital Medical School, data from historically biased processes will generate similarly biased predictions. In this way predictive policing, while supposedly neutral, overemphasizes the criminality of black neighborhoods and intensifies the police presence around people and places that are already unfairly targeted.

Yet because this outcome is based on data, dispatching police in this manner is typically seen as an objective rather than political decision. When the recommendations of predictive policing are taken at face value, what was once a racially motivated decision by a police force to crack down in certain neighborhoods becomes an objective response based on science. Thus, writes the data ethicist Jacob Metcalf, the original "value decision" of whom to arrest "becomes naturalized through the black-box 'objectivity' of the algorithm."[30] Such naturalization may create a pernicious feedback loop that justifies and perpetuates systemic biases.

An analysis in Oakland by the Human Rights Data Analysis Group demonstrates how predictive policing can lead to these disparities. Although local public health estimates suggest that drug crimes are ubiquitous in Oakland, the study found that "drug arrests tend to only occur in very specific locations—the police data appear to disproportionately represent crimes committed in areas with higher populations of non-white and low-income residents."[31] The study's authors developed an algorithm, based on PredPol's methods, to determine what impacts predictive policing could have. They concluded that if the Oakland Police had used PredPol, "targeted policing would have been dispatched almost exclusively to lower income, minority neighborhoods."[32]

But perhaps some technical mechanisms can be employed to avoid biased predictions. Jeremy Heffner is the product manager of HunchLab, a competitor of PredPol that attempts to answer the obvious question: if the biases reflected in crime data make predictive policing discriminatory, is there any way to make the models fair? "One thing that PredPol puts forward as the strength of their approach is that they're only using crime data, so therefore there's no possibility for bias, which just makes no sense to me," Heffner says. "If there's any bias in the systems, it's because of the crime data itself."[33]

Heffner has taken numerous steps to prevent these biases from appearing in HunchLab. His primary focus is limiting the use of data that is influenced by an officer's discretion. Heffner provides the following example: "When an officer is going down the street, they are not generating new reports of robberies or homicides, but they might be generating new reports of vandalism or jaywalking." This means that data about homicides is less likely to reflect biases—and therefore generate biased predictions—than is data about vandalism. Moreover, HunchLab incorporates numerous other

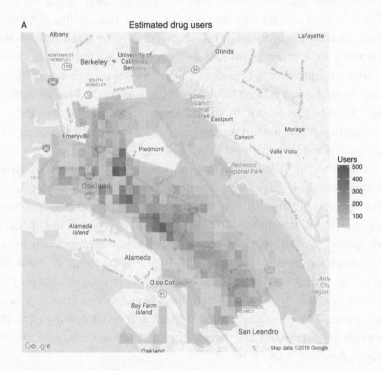

Figure 4.1
A demonstration in Oakland of how predictive police algorithms can perpetuate the biases embedded within crime data. Although (a) estimates of drug use span the city, (b) the Oakland Police disproportionately made drug arrests in low-income and minority neighborhoods. In turn, (c) a typical predictive policing algorithm would dispatch police almost solely to those parts of the city.
Source: Kristian Lum and William Isaac, "To Predict and Serve?," *Significance* 13, no. 5 (2016): 17–18.

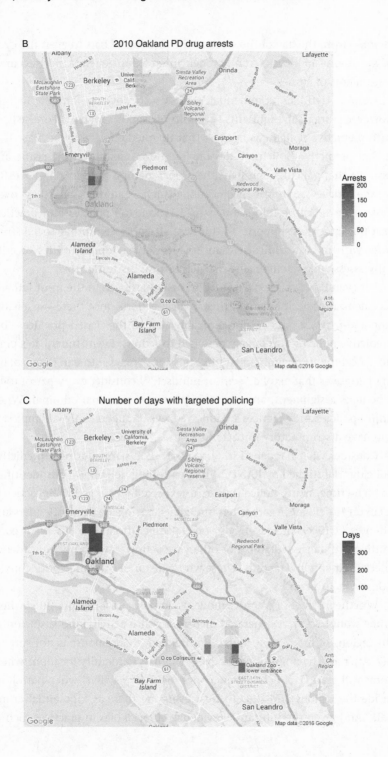

factors—from the day of the week to the location of bars to even the cycle of the moon—in a "risk terrain model" intended to generate more accurate assessments of crime.

Despite Heffner's commendable efforts to develop fair predictions, however, HunchLab highlights the limits of "smart" policing. For the issue with predictive policing is not just that the predictions may be biased—it is that predictive policing relies on traditional definitions of crime and assumes that policing represents the proper method to address it. Focusing on the models' technical specifications (such as accuracy and bias) over-looks an even more important consideration: the policies and practices that the algorithm supports. In this way, attempts to improve social struc-tures with mere technical enhancements subvert opportunities to criti-cally assess and systematically reform political institutions.

For even if police are dispatched to neighborhoods in the most fair and race-neutral possible manner, their typical actions once there—suspicion, stop-and-frisks, arrests—are inextricably tied to the biased practices that predictive policing was largely designed to redress. Even HunchLab's prac-tice of calling its recommendations "missions" plays into dangerous polic-ing narratives that extol a "warrior mindset,"[34] consider every patrol to be a perilous assignment, and view everyone as a potential criminal. When unjust policies and practices are followed, even a superficially fair approach will have discriminatory impacts.

Consider what happened in Shreveport, Louisiana, during a predictive policing trial studied by RAND. When patrolling neighborhoods identified as high-crime, many police officers unexpectedly changed their tactics to focus on "intelligence gathering through leveraging low-level offenders and offenses." Officers increasingly stopped people whom they observed "com-mitting ordinance violations or otherwise acting suspiciously" in order to check their criminal records. Those whose histories contained prior convic-tions were arrested.[35]

Whether or not Shreveport's model accurately and fairly identified where crime would occur, it generated increased police activity and suspicion in the regions of interest. Although unintended, this response is not surpris-ing. After all, the point of predictive policing is to identify locations where crime will occur. Doing so primes police to be "hyper alert" when patrolling inside the regions and thus to treat everyone there as a potential crimi-nal.[36] And given the substantial evidence of racial bias in practices such as stop-and-frisk,[37] it is not hard to imagine that the people whom police stop for committing violations or acting suspiciously will mainly be young men of color, thereby increasing both incarceration rates and conflict between police and communities.

Here we see the interplay between predictions and politics: whether or not predictive policing algorithms accurately and fairly identify high-crime locations, they do not dictate what actions to take in response. Governments *choose* to give responsibility for dealing with most forms of social disorder to the police. Police *choose* to go into these neighborhoods with heightened suspicion and a warrior mindset. Thus, the seemingly technical decisions about how to develop and use an algorithm are necessarily intertwined with the clearly political decisions about the kind of society we want to inhabit. Just as it is necessary to assess and reimagine policing generally, it is equally necessary to assess and reimagine the role of algorithms in policing. For if cities truly know where crime will occur, why not work with that community and with potential victims to improve those neighborhoods with social services?[38] Why is the only response to send in police to observe the crime and punish the offenders?

* * *

Proponents of "smart" policing are typically so focused on optimizing existing practices that they are unable to answer—or even ask—questions about what should be done with the predictions that are made. Policing is not the only or the most effective way to curb crime and aid communities—in fact, as the police scholar David Bayley explains, "one of the best kept secrets of modern life" is that "[t]he police do not prevent crime."[39] For example, a 2017 study found that proactive policing "may inadvertently contribute to serious criminal activity" and "curtailing proactive policing can reduce major crime," suggesting that one of the most common (and discriminatory) police practices does not even achieve its stated purpose of reducing crime.[40]

Although police possess means and powers to deter and punish certain criminal activity, they are ill-equipped to take on the full range of issues with which they are increasingly required to deal: homelessness, mental health and drug crises, isolated neighborhoods with poor education and limited job opportunities. These issues would be better addressed by alternative interventions.

"I don't think anyone, in the abstract, has a problem with figuring out where crime is and responding to it," says the ACLU's John Chasnoff. "But what's the appropriate response? The assumption is: we predicted crime here, and you send in police. But what if you used that data and sent in resources?"[41]

That's what happened in Johnson County, Kansas—a county within the Kansas City Metro Region. The story starts decades ago, in 1993, when the sheriff, chief judge, and district attorney each sought funding from the county manager for new record management systems. The county manager was unwilling to pay for three separate versions of almost identical software, so he told them to come up with one record management system that could meet all of their needs.[42]

The county manager got his wish: the group worked together and created a single, integrated information management system that combines data about every criminal case, from booking through the conclusion of probation, in one place. In 2007, the county also integrated its human services data into the same system.

In addition, Johnson County has spent years developing policies that prioritize coordinated treatment for individuals suffering from mental illness. In 2008, it formed a cross-governmental Criminal Justice Advisory Council to assess the local criminal justice system and identify gaps in social services. One of the council's first initiatives was hiring mental health professionals whose job was to help police respond to incidents that involve mental health issues. After this program was successfully launched in one of the county's cities in 2011, slightly reducing jail bookings of the mentally ill and increasing referrals to services more than thirtyfold, it expanded to another city in 2013.[43] Soon thereafter, every city in Johnson County had made appropriations in its budget for a qualified mental health professional to be embedded in its police department.[44]

In 2015, the success of these efforts in Johnson County attracted the attention of the White House. Lynn Overmann, then the senior advisor to the U.S. chief technology officer, was pulling together a select cohort of jurisdictions for a fledgling "Data-Driven Justice Initiative" and wanted Johnson County to be involved. The initiative's goal was to use data to help address a crisis in the criminal justice system: the shocking frequency with which people with unaddressed mental illnesses are locked up in local jails for committing minor, nonviolent offenses. Two-thirds of inmates suffer from mental illness, two-thirds have a substance abuse disorder, and almost half suffer from chronic health problems.[45] Jailing them costs billions of dollars every year.

A major reason why so many people with mental illness end up in jail is that most communities lack the services and coordination necessary to address this population's multiple vulnerabilities, which can include drug addiction and homelessness. Even though many agencies devote resources to this population, they do so in piecemeal ways that fail to sufficiently aid and stabilize individuals.[46] As a result, Overmann explains, "America's largest mental health facilities are often our local jails."[47] But relying on police and jails is an inadequate approach that merely places a punitive Band-Aid on systemic and complex issues. As one county sheriff says, "These are not issues we can arrest or incarcerate our way out of."[48]

Overmann has observed firsthand how communities are failing their most vulnerable residents. She began her career as a public defender in Miami, where she "saw from the inside how ill-equipped the criminal justice system is to help people with mental illness." Overmann reports that although many of her clients suffered from mental health issues, "they lacked access to required mental health services. As a result, these clients often spent weeks or even months in jail."[49] Conditions for prisoners suffering from mental illness in Miami were so bad that in 2011 the Department of Justice declared the situation "inhumane and unconstitutional."[50] The lesson for the young Overmann was clear: "The system was broken."[51]

In response to these issues, Miami revolutionized how it treats people with mental health issues. After discovering that fewer than 100 people with serious mental illness accounted for nearly $14 million in services over four years, the Miami-Dade Police Department trained police officers and 911 dispatchers to de-escalate encounters with people suffering from mental illness.[52] Because of the police department's focus on humane treatment and diversion away from jail and into social services, the local jail population fell by 40 percent, a decrease so steep that the county saved $12 million per year and was able to close an entire jail facility.[53]

Overmann carried her experience in Miami to the White House Office of Science and Technology, where she had the platform to help communities across the country divert low-level offenders with mental illness away from the criminal justice system and into treatment. Her goal was to proactively provide coordinated social services to those with mental health problems and criminal records before they ever came into further contact with the criminal justice system.

Overmann knew that success would be largely predicated on having accurate and functional data. In theory, determining the overlap between people who have been arrested and people who receive mental health treatment is easy: just combine datasets and see which names appear in both. But in reality, even this first step—merging datasets from separate components of local government—is incredibly challenging. The data that municipal agencies collect is typically used for internal administrative purposes (e.g., tracking building permits and dispatching ambulances) rather than analysis; each department is focused intensely on its own specific responsibilities, paying little attention to sharing datasets across departments. Each department's records are therefore isolated in individual silos, created and maintained in whatever form best supports its particular objectives. These bureaucratic barriers to data integration create significant blind spots for agencies trying to coordinate their services: those running criminal justice systems "don't know how many people screen positive for mental illness," and behavioral health clinicians "never know if our clients are in jail."[54] As a result, agencies are unable to meet the needs of this vulnerable population.

The importance of integrated datasets is what made Johnson County such an attractive test-bed for the Data-Driven Justice Initiative. Because of its decisions over the past several decades, first to create a unified criminal justice information management system and then to integrate human services data, Johnson County possessed the vital information that most jurisdictions lacked.

In 2016, Johnson County partnered with the University of Chicago's Data Science for Social Good program on an ambitious project: to identify which individuals suffering from mental health and medical issues would be arrested in the following year. With this information, Johnson County could provide proactive social services that would enable someone with mental illness to avoid coming into any further contact with the criminal justice system. The goal was not just to divert people from jail but to prevent them from ever reaching a crisis that requires diversion.

Toward this end, the team at UChicago developed a machine learning model. Johnson County's data contained detailed records for 127,000 people. The data scientists consulted with Johnson County to develop 252 features (including age, history of criminal charges, and number of times enrolled in mental health programs in the past year) that could help predict future arrests. They also categorized everyone in the data on the basis of whether they had recently been arrested. Using this labeled training data, the team developed a predictive model to determine each person's likelihood to be arrested in the following year.[55]

The algorithm identified several trends that indicate when someone with mental illness is likely to be arrested. Most notably, the highest-risk individuals had long gaps between their interactions with mental health services, suggesting that dropping out of social services prematurely greatly increases someone's risk of coming into contact with the criminal justice system. Drawing on these insights, the model could automatically detect people who were following this trajectory and help Johnson County intervene before they fell through the cracks.

A retrospective analysis showed just how much the model could aid those in Johnson County. Among the 200 people who were identified as having the highest risk to be arrested in 2015, 102 went to jail in that year. If Johnson County had proactively reached out to this high-risk population, half of them might have been kept out of jail. The impacts of this predictive approach could be profound: preventing the bookings of these 102 people would have spared them almost eighteen years of cumulative jail time and, as an added benefit, would have saved the county about $250,000.[56]

Steve Yoder, a data specialist in Johnson County who worked with UChicago on the project, remembers being skeptical at first that there were so many people being jailed on a regular basis because of mental health issues. But when he started looking through the data, he was struck in particular by one person on the model's list who had been booked into jail six times in the past six months.

"For those of us who don't experience this on a day-to-day basis it's just hard to imagine," Yoder explains. But after double-checking the numbers, he recognized how severe the problem truly is. "This is real. There's a person behind that. And, wow, there's some crisis going on here that really needs to be addressed."[57]

As Johnson County moves toward implementing this model to guide social service outreach, it is working with the University of Chicago to predict high-risk individuals on a monthly, rather than yearly, basis. Meanwhile, the Data-Driven Justice Initiative continues to expand. At the end of the Obama administration, the initiative found a new home in the Laura and John Arnold Foundation.[58] Across the country, from Los Angeles to Salt Lake County to New Orleans to Hartford, more than 150 jurisdictions are combining previously disparate datasets to provide more proactive and effective social services.[59]

"I'm a believer in this," proclaims Yoder. "I truly think that we haven't even scratched the surface of the data."

* * *

The work in Johnson County is remarkable in part because it starts from the understanding, as the county's criminal justice coordinator Robert Sullivan puts it, that "people have all kinds of complexity in their lives. Some of their interactions with the criminal justice system are due to those conditions." With this recognition comes the agency to make substantive change: rather than simply treating predicted outcomes as preordained, Johnson County uses predictions to inform preventive interventions that alter those outcomes. "We don't want you to ever get to the point of having an interaction with any component of the criminal justice system," Sullivan says. "That's why we're so excited about this predictive piece."

Johnson County's perspective stands in stark contrast to the view of the world through tech goggles: the only possible social change is to make policing more efficient by using data and algorithms. As PredPol explains on its website, its core mission is to help police departments "allocate the limited resources they do have more effectively."[60] By this logic, a smart city is one that adds technology to traditional practices in order to catch criminals and lower crime rates.

Yet creating a just city means more than merely optimizing typical police practices with efficient crime prevention in mind. For example, policing is a job with numerous, often conflicting goals—it cannot be boiled down to a number or formula. "It's hard to measure success for a police force in a truly and meaningfully holistic fashion," says David Robinson. "Police are trying to maintain legitimacy, they're trying to deter crime, they're trying to investigate crimes that happened, they're trying to create social order without manufacturing indignity in the lives of the people that they supervise. The crime rate is a very poor substitute for having a comprehensive metric of success for a police department."

Failure to incorporate the complexities of policing into predictive models can be disastrous. Just as an algorithm that optimizes traffic flow will overlook the needs of pedestrians, a predictive policing model that optimizes for reducing crime rates will ignore the other responsibilities of police as well as other goals for the community. And just as a spam filter that focuses on catching phishing emails will struggle to capture other types of spam, a crime forecast that focuses on drug crimes rather than white-collar crimes will unfairly place a big, fat target on minority neighborhoods. It is only by starting with a comprehensive and compassionate understanding of what factors lead to contact with the criminal justice system and what tactics can be used in response, as Johnson County did, that algorithms can truly help generate a more just city.

Instead of conceiving more holistic approaches that capture the complexity of the world, however, engineers tend to adopt visions of society that fit the simple presumptions within their models. Take Richard Berk, a professor of statistics and criminology at the University of Pennsylvania who has spent his career using data to analyze crime. Several of his projects involve helping judges decide which inmates should be released on parole by predicting who is likely to recidivate. Berk describes the task in vivid terms: "We have Darth Vaders and Luke Skywalkers, but we don't know which is which."[61] The goal is to distinguish Vaders from Skywalkers.

Although this description helps explain how the algorithm works, it provides a stunning oversimplification of society. Have you ever met a Darth Vader? Who among us is a Luke Skywalker? The world cannot be broken down into people who want to destroy the universe and those who risk their lives to save it. Unwittingly, Berk's chosen analogy highlights the fallacies of such simplistic thinking. As one critic writes, "Berk must not have watched the entire 'Star Wars' saga. Darth Vader wasn't an unimpeachably evil individual. At one point he was an innocent little boy who grew up in some dire circumstances."[62] Rather than question why people make certain decisions or end up in particular situations—and attempt to push them toward positive outcomes—Berk presumes that people are fundamentally either good or bad, and that our task is simply to determine whom to punish. Apparently all we can do is follow the binary representations defined by the algorithm.

One of Berk's most ambitious efforts is to predict whether newborn babies will commit a crime before turning eighteen, from information such as where that baby lives and who its parents are.[63] He is starting in Norway, but if the same approach is taken in the United States, there is little doubt that a machine learning model could distinguish with reasonable accuracy between people who will be arrested and those that will not. After all, a government report estimated that of male babies born in 2001, one of out every three blacks, compared to only one out of every seventeen whites, would go to prison at some point during his life.[64] Given those stark statistics, we don't need cutting-edge algorithms to predict who will be arrested.

But just because we can predict a certain outcome does not mean we should consider that outcome to be inevitable or just. That a model could predict a baby's future criminality reflects the vast inequalities of justice and opportunity in society, not the inherent nature of certain people. In just the last century, African Americans have, among many injustices, been excluded from government programs that provided loans for education and housing and been funneled into prisons through the war on drugs.[65] The vast disparities in education, wealth, and crime that have resulted from these actions are not inevitable but socially constructed. To suggest that an algorithm can identify future criminals at birth is thus to accept the status quo as the natural and proper state of society, in effect labeling fights for equity and social justice as unnecessary.

A 2012 advertisement for IBM's Domain Awareness System portrays a similar perspective. The commercial follows two white men—the proverbial cop and robber—driving through city streets at night. The police officer provides a voiceover that begins as follows: "I used to think my job was all about arrests. Chasing bad guys. Now I see my work differently. We analyze crime data, spot patterns, and figure out where to send patrols." Relying on the advice of a computer in his police car, the officer reaches a convenience store just in time to thwart the would-be thief.[66]

Although it tells an appealing story, IBM's ad demonstrates how predictive policing software both relies on and perpetuates simplistic notions of policing and crime. The officer's first two statements set up the rules of society: there are "bad guys" who commit crime and police (the implied "good guys") whose job it is to arrest them. This story presents another Luke Skywalker versus Darth Vader scene, with no backstory (for apparently none is needed) to explain how each person came to their present roles. In this way, in addition to completely exaggerating what algorithms are capable of—no system can predict crime at scale with anywhere near the level of precision depicted—IBM's ad ignores all of the social and political dynamics that underlie crime and policing. The society portrayed in this vignette has no poverty, no segregation, no stop-and-frisk—in fact, because every character is white, it has no racial dynamics at all. We are left with a facile and pernicious conclusion: because of the presence of "bad guys," crime is an inevitable phenomenon that can be prevented only by police who possess the necessary information.

This is the pernicious logic that the tech goggles cycle reinforces. First, we perceive policing as a purely technical problem of deploying officers to prevent crime. Rather than evaluate whether current police practices are well suited to addressing social disorder, we deploy predictive policing algorithms to slightly adjust police operations. Because tech goggles create a mirage of objectivity around data and algorithms, technological approaches like predictive policing are perceived as value-neutral responses to social problems. And in order to justify myopic models that do not—for they cannot—capture the full complexity of society, we adapt our social theories to match the world that the models depict. Police departments and courts become further entrenched in their view of the population as good or evil and of incarceration as the only response to crime.

When deployed within this framework, machine learning will be an ineffectual (at best) or counterproductive (at worst) tool for social justice. Consider once again the Cambridge Police Department's housebreak pattern detection algorithm, which was motivated by precisely the kind of crime prevention that IBM's ad portrays and which represents perhaps the best-case scenario for what crime-predicting software in the hands of police can look like. Better investigation and prevention of burglaries could benefit many. The CPD relied on data solely about housebreaks, information that is relatively reliably reported to and recorded by police. Furthermore, their algorithm is primarily intended for retroactive investigations and targeted pattern detection, as an explicit counter to software such as PredPol that directs police where to patrol to proactively prevent crime. "It's absolutely the wrong approach to come into neighborhoods and stop everyone," says Dan Wagner. "That's been the problem in policing and problem with these tools."

But even the CPD's work, like every other attempt at predictive policing, suffers from a gaping divide between the problem being solved and the problem that needs solving. Owing to their focus on technology, many believe that the issues of policing stem from poor information about when and where crime will occur in the future. This is a problem that (at least in principle) new technology can solve. But as Alex Vitale argues in *The End of Policing*, "The problem is not police training, police diversity, or police methods. . . . The problem is policing itself." Tracing the history of policing from its roots to the present day, Vitale concludes: "American police function, despite whatever good intentions they have, as a tool for managing deeply entrenched inequalities in a way that systematically produces injustices for the poor, socially marginal, and nonwhite."[67] Cities do not need to embrace new technology so that they can improve police capabilities—they need to fundamentally reconceptualize the roles, practices, and priorities of police.

In the hands of police, even algorithms intended for unbiased and nonpunitive purposes are likely to be warped or abused. For whatever its underlying capabilities, every technology is shaped by the people and institutions that wield it. Unless cities alter the police's core functions and values, use by police of even the most fair and accurate algorithms is likely to enhance discriminatory and unjust outcomes.

In Chicago, for example, an algorithm conceived to reduce violence was perverted—through police control—into a tool for surveillance and criminalization. Drawing on his research regarding how gun violence clusters in social networks,[68] the sociologist Andrew Papachristos urged that social service organizations identify people facing the highest risk of being shot in order to prevent future violence and mitigate its impacts. Building on these insights, the Chicago Police Department developed an algorithm to identify the people most likely to be involved in gun violence. And although the original stated intention for this "Strategic Subjects List" (SSL) was to prevent violence, it has largely been used as a surveillance tool that many believe disproportionately targets people of color.[69] A RAND evaluation concluded that the SSL "does not appear to have been successful in reducing gun violence"; instead, "the individuals on the SSL were considered to be 'persons of interest' to the [Chicago Police Department]" and were more likely to be arrested.[70] Even Papachristos has criticized this application of his research, writing in the *Chicago Tribune* that "one of the inherent dangers of police-led initiatives is that, at some level, any such efforts will become offender-focused."[71]

Smart Enough Cities must take machine learning out of the hands of police and develop nonpunitive and rehabilitative approaches to address social disorder, along the lines of the Data-Driven Justice Initiative. Wagner advocates for precisely these transformations. "The criminal justice system ends up putting people in jail, and that's not an effective place to treat them. They need mental health or substance abuse services," he says.

And although Wagner is confident that algorithms have a role to play, he argues that we must think more critically about how to use them: "I think there's value in using social networks to identify people who are at risk of being involved in a shooting, but Chicago fell flat in implementing the Strategic Subjects List to increase surveillance of suspects as opposed to truly trying to prevent that person from pulling the trigger or being shot. If they had used that same tool and better partnered with the community, it would have been very different."

* * *

Many of the advances promised in smart cities rely on data analysis and machine learning algorithms—presented as providing universal benefits— yet these techniques are unable to transcend historical or current politics.

First, the data used to develop these models does not represent unassailable truths; instead, the data embeds information about socially produced outcomes and is shaped by reporting and collection practices. As the examples of 311 and police data indicate, often what we believe to be data about one thing (potholes and crime) is in fact data about something quite different (service requesting inclinations and police activity). Given that machine learning relies on historical data, we should be critical of what predictive algorithms actually forecast and hesitant about using them to direct municipal operations.

More fundamental than biases within data are the politics embedded within the algorithms. For although designing algorithms appears to be a technical task, the choices made can have vast social and political impacts. All too often, algorithms that promise efficiency as a neutral good reflect the priorities of existing institutions and power structures. In privileging police efficiency in reducing crime rates over alternative goals such as improving neighborhood welfare with social services, supposedly neutral models further entrench the role of police as the appropriate response to social disorder—a political decision if there ever was one. In that sense, predictive policing is likely to have discriminatory impacts not just because the algorithms may themselves be biased but also because they are deployed to grease the wheels of an already discriminatory system.

Rather than rush to adopt machine learning, we must ask: What goals should we pursue with the aid of predictive algorithms? How should we act in response to the predictions that are generated? How can we alter social and political conditions so that the problem we want to predict simply occurs at lower rates? Not every application of machine learning is inevitably biased or malicious or useless, but achieving benefits from machine learning requires that we debate—in political rather than technical terms—how to design algorithms and what they should be deployed to accomplish.

While the criminal justice system (not to mention every other aspect of municipal governance) has always involved contentious and complex political decisions, the particular danger of using technology to make these decisions is that we will misinterpret them as technical problems that do not require political deliberation. And by treating technology as the only variable, tech goggles blind us to the full possibilities to reform the policies and practices that technology purports to improve. When predictive policing gets hailed as the new and scientific approach to policing, it distracts us from the hard choices that must be made about what police should prioritize and what their role in society should be. Thus, says Andrew Ferguson, "Predictive policing systems offer a way seemingly to turn the page on past abuses, while still legitimizing existing practices."[72]

As traditional practices are cloaked in the futuristic sheen of algorithms, they are made to appear more innovative and attractive than they truly are. Looking through tech goggles, we mischaracterize applying new technology to the same old practices as progress. But there are no easy technical fixes for systemic police discrimination and the debilitation of social services: more substantive reforms are required. Johnson County's efforts were effective not because it discovered a new, infallible algorithm to optimize existing police practices, but because it developed strategies to address mental health issues, created the data infrastructure necessary to inform those interventions, and devoted sufficient resources to make the interventions effective. Unlike those who leap at the quick-fix solution promised by predictive policing, Robert Sullivan emphasizes that improving the criminal justice and mental health systems required an "incremental step-by-step process through the years." As chapter 6 will discuss more thoroughly, social progress that appears driven by technology actually relies on precisely these sorts of long-term planning efforts and nontechnical policy reforms.

But first, the next chapter will further explore how using data and algorithms in government is primarily a political rather than technical project, examining how cities should responsibly govern these technologies to ensure that their use promotes democracy and equity.

5   The Responsible City: Avoiding Technology's Undemocratic Social Contracts

Throughout the book, we have explored how technology's social impacts are influenced by far more than simply its capabilities: social and political conditions constrain the outcomes that technology can generate, people can wield similar technologies to achieve distinct outcomes, and machine learning models derive insights from data that reflects a legacy of bias and discrimination. By overlooking these factors, tech goggles encourage the adoption of smart city technologies that generate unintended and undesirable social consequences.

This chapter introduces one more component to the discussion: the technical and political arrangements of technology, also known as its architecture. Questions related to architecture go beyond a technology's purpose and output—What should this algorithm do? Is it accurate?—and encompass its structure: By what means should this technology achieve its goals? Who should control it? How should we pay for it?

The answers to these questions can have significant ramifications. The ways in which technology structures social and political relations can be even more consequential than the explicit function that it serves. When developing and adopting new technology, we must therefore "examine the social contract implied by building that [technological] system in a particular form," asserts Langdon Winner. "[A]s our society adopts one sociotechnical system after another it answers some of the most important questions that political philosophers have ever asked about the proper order of human affairs," writes Winner. "Just as Plato and Aristotle posed the question, What is the best form of political society?," today we must ask, "What forms of technology are compatible with the kind of society we want to build?"[1]

Whether we recognize it or not, the technologies we implement in cities today will play a significant role in defining the social contract of the next century. And as it currently stands, the architecture of the smart city is a fundamentally undemocratic one: many technologies operate by collecting unchecked data about individuals and using opaque, often proprietary, algorithms to make life-altering decisions. In the process, they create massive information and power asymmetries that favor governments and companies over those they track and analyze, breeding impotence and subjugation within society. In this way, the smart city is a covert tool for increasing surveillance, corporate profits, and social control.

City governments eager to take advantage of new technologies must act as responsible gatekeepers and public stewards in structuring their technology to protect equity and fundamental rights. The Smart Enough City need not accept the unsavory compromises of smart cities—more democratic approaches to utilizing technology are possible.

* * *

Equitable access to high-speed internet has become a necessary staple of democratic society. Without internet access, it is difficult if not impossible to apply for jobs, access healthcare, and connect with other people. Yet because of the high price of internet subscriptions, many low-income individuals and families are unable to afford reliable broadband access. In Detroit, for example, 40 percent of residents lack broadband.[2] In New York City, that number is 23 percent.[3]

In 2016, it appeared that New York City had found the solution to this digital divide that could be a model for every city: LinkNYC, a program to provide free public Wi-Fi via more than 7,500 internet-connected kiosks placed throughout the city (as of November 2018, 1,742 had been installed).[4] "LinkNYC brings us a couple steps closer to our goal of leveling the playing field and providing every New Yorker with access to the most important tool of the 21st century," proclaimed Mayor Bill de Blasio at the program's launch. Perhaps most amazingly, providing this service will not cost the government a cent—in fact, NYC expects the program to bring in more than $500 million in revenue to the city by 2025.[5]

This appears, like many smart city technologies, to be a benevolent technical solution for an important social problem. But under the surface, where LinkNYC's architecture resides, lurks a more insidious reality.

The benefits and finances of LinkNYC sound too good to be true. So how is the program funded? The kiosks are owned and operated by a consortium of companies that includes investment and leadership from Sidewalk Labs, a subsidiary of Alphabet (Google's parent company), which plans to pay for the initiative by collecting and monetizing data about everyone who uses the service. As Sidewalk founder and CEO Dan Doctoroff told a public audience in 2016, the company expects to "make a lot of money from this."[6]

LinkNYC kiosks are equipped with sensors that gather an enormous quantity of data about every device that connects to the Wi-Fi network: not just its location and operating system but also its MAC address (a device's unique identifier that helps it connect to the internet).[7] Sidewalk Labs claims that this data is merely "Technical Information," in contrast to the "Personally Identifiable Information" it collects such as your name and email (which are required to register for network access).[8] This distinction follows traditional standards of privacy, which focus on the presence of personally identifiable information (PII)—features such as name, address, and Social Security number that, on their own, identify individuals. Data containing PII is considered sensitive, while data without PII is not.[9]

To the human eye, this appears to be a sensible distinction. After all, MAC addresses are twelve-digit alphanumeric strings that look like indecipherable, machine-processing gobbledygook. But just because data does not contain names and is difficult to decipher does not mean it lacks information about people. Yes, one single data point—a phone's MAC address at a particular place at a particular time—is unlikely to reveal someone's identity or anything sensitive about them. But when millions of data points are collected and combined with modern analysis techniques, such data can be used to track people's movements and infer intimate details of their lives.

This data is so sensitive in aggregate, despite each record seeming so benign in isolation, because human behavior is incredibly unique. Data collected on a massive scale captures these particularities. Research led by the computer scientist Yves-Alexandre de Montjoye demonstrated this phenomenon by analyzing two datasets that contained mobile phone location traces and credit card transactions about more than one million people.[10] Even though both datasets lacked PII—they included just anonymous personal IDs (à la MAC addresses), locations, and times—de Montjoye revealed that it was possible to identify individuals and learn about their behavior. Remarkably, more than 90 percent of people could be uniquely

shop	user_id	time	price	price_bin
🥾	7abc1a23	09/23	$97.30	$49 – $146
🍓	7abc1a23	09/23	$15.13	$5 – $16
🛒	3092fc10	09/23	$43.78	$16 – $49
🍔	7abc1a23	09/23	$4.33	$2 – $5
🏊	4c7af72a	09/23	$12.29	$5 – $16
🍔	89c0829c	09/24	$3.66	$2 – $5
✕	7abc1a23	09/24	$35.81	$16 – $49

Figure 5.1

A stylized example of behavioral data that includes no personally identifiable information (PII) but nonetheless contains information about individuals. By examining all of the records pertaining to a given person, it is possible to infer their behavior. The map above traces the activity of the person represented by user_id=7abc1a23.

Source: Yves-Alexandre de Montjoye, Laura Radaelli, Vivek Kumar Singh, and Alex "Sandy" Pentland, "Unique in the Shopping Mall: On the Reidentifiability of Credit Card Metadata," *Science* 347, no. 6221 (2015): 537. Reprinted with permission from AAAS.

identified with just four data points of where they have been and when they were there.

Although de Montjoye's analysis shed new light on the privacy risks of granular behavioral data, it was not the first demonstration of how supposedly anonymous data can reveal a great deal about people. That came in 1997, when Massachusetts Governor William Weld released state employee medical records for research purposes, promising that the information was anonymous. A few days later, however, Weld received a letter: it contained his own health records, culled from the released data.[11] The envelope came from Latanya Sweeney, then a graduate student at the Massachusetts Institute of Technology, who had identified Weld's file by linking the medical records with publicly available voter lists via information contained in both datasets (such as birth date).[12]

Many other datasets come with similar risks of reidentification. When New York City released supposedly anonymous data about every local taxi trip in 2013, a data scientist analyzed patterns of where those rides began and ended to identify the names of several patrons of a Manhattan strip club.[13] The same method could be used to learn who prays at mosques, works late nights, visits gay bars, or undergoes chemotherapy. Another data scientist used similar data about trips on London's bike share program to reveal the travel patterns of several individuals, deducing where they live and work.[14]

But the danger is not merely that one's identity and behavior can be revealed from seemingly anonymous data—when data is combined with artificial intelligence, it is possible to infer a great deal of personal information that is not explicitly contained in a dataset. With detailed information about where you have been, for instance, machine learning algorithms can predict whom you know and where you will go next.[15] Algorithms can detect whether someone is depressed on the basis of photos she posts on Instagram.[16] Data about seemingly routine behaviors such as Facebook Likes can reveal sexual identity, race, political affiliation, and even whether one's parents are married.[17]

The potential for algorithms to identify and learn about people using supposedly anonymous and benign data highlights the remarkable invasion of public privacy presented by LinkNYC and reveals the definitional trick that enables it: Sidewalk Labs's "Technical Information" is presented as being anonymous, but it is in fact far more sensitive than the "Personally Identifiable Information" that the company so magnanimously promises to safeguard. In other words, says one privacy lawyer, the LinkNYC privacy policy is designed "to make you believe that something is being promised, when actually it lets them do anything they want."[18] The motivation is profit: the more detailed the data, the better Sidewalk Labs can monetize it.

Recognizing these privacy risks, many New Yorkers have raised concerns about LinkNYC.[19] The New York Civil Liberties Union's executive director has argued, "Free public Wi-Fi can be an invaluable resource for this city, but New Yorkers need to know there are too many strings attached."[20]

In *The Digital Person*, the privacy scholar Daniel Solove highlights two fundamental ways in which such widespread data collection and concomitant knowledge threaten society. The most salient concern is widespread surveillance, as governments and companies are allowed to watch your every action, expose secrets, or even catch everyone who runs a red light. Such fears tap into deep-seated cultural notions about privacy that are drawn from images of Big Brother, the totalitarian government in George Orwell's 1949 novel *1984*. By observing the most intimate details of everyone's lives and punishing even the slightest dissent, Big Brother controls the society's behavior. Following Orwell's influence, writes Solove, we typically conceive of privacy following the "secrecy paradigm": the idea that privacy is invaded when one's secrets are observed or exposed, leading people to self-censor (via "chilling effects") or suffer the consequences.[21]

1984-inspired fears capture a great deal of why privacy is essential for maintaining civil liberties. "The government has a long history of spying on activists in an effort to dampen dissent and we see those same actions present today, in an age where most people have substantial digital footprints," explains the activist DeRay McKesson.[22] The FBI in the 1960s, for example, focused surveillance on civil rights activists such as Martin Luther King, Jr., in order to intimidate and harass them.[23] This history appears to be repeating itself, as federal and local law enforcement officials have been tracking the identities and activities of those protesting police violence in the Black Lives Matter movement.[24]

Yet Big Brother cannot explain every risk of declining privacy. As we have already seen, a significant portion of data collection today relies on information that is neither secret, illegal, nor embarrassing—in fact, many individual data points appear meaningless and anonymous. The secrecy paradigm thus fails to explain the harms of someone's bike share trips or Facebook Likes being collected, aggregated, and analyzed. As Solove explains, nowadays many uses of data "aim not at suppressing individuality but at studying it and exploiting it."[25]

Solove likens much of today's collection and use of data to the themes of another twentieth-century novel: Franz Kafka's 1925 *The Trial*. The book's protagonist, Josef K., wakes up on his thirtieth birthday to find two men in his room declaring that he is under arrest. He is given no indication of what he has done or what agency is arresting him. The novel depicts Josef's unsuccessful attempts to uncover the identity of the mysterious court and what data it possesses about him. He is murdered by the court's agents on his thirty-first birthday without ever having learned their true nature. "*The Trial* captures an individual's sense of helplessness, frustration, and vulnerability when a large bureaucratic organization has control over a vast dossier of details about one's life," Solove explains. In doing so, it "captures the scope, nature, and effects of the type of power relationship created by databases."[26]

Just like Josef, people today have little knowledge of or control over what personal data is collected, who owns it, and how they exploit it. As more data is gathered and used by governments and companies, privacy becomes defined less by the secrets that any single piece of information reveals and increasingly by the inferences that large amounts of relatively nonsensitive data make possible—and the power that those inferences grant. For example, Facebook can calibrate its News Feed algorithm to influence a user's mood and likelihood to vote.[27] OkCupid can alter profile match scores to affect certain people's chances of getting dates.[28] Healthcare services can deny coverage if they learn that someone recently visited websites associated with having cancer.[29]

Although data collection touches everyone, the most severe impacts of diminishing privacy are suffered by the poor and minorities. Despite being more concerned than their more well-off counterparts about privacy, most lower-income individuals lack the knowledge of privacy settings and policies to sufficiently reduce the extent to which they are tracked.[30] And given that activists in racial justice movements like Black Lives Matter are targeted for surveillance and undocumented immigrants face deportation, minorities are prone to suffer the most damaging consequences of being identified and tracked by the government.

Moreover, those with the lowest socioeconomic status often have no choice but to accept government oversight in exchange for social services. Welfare offices use electronic benefits transfer (EBT) cards to monitor recipients' behavior in increasingly pervasive and intimate ways. As the political scientist Virginia Eubanks explains, these technological practices "significantly limit clients' autonomy, opportunity, and mobility."[31]

This suffocating oversight has long been a feature of government services. In his 2001 book *Overseers of the Poor*, the political scientist John Gilliom documents how welfare recipients are closely monitored by the government through endless paperwork and meetings with fraud control agents to ensure that they are eligible for services and comply with the many requirements. The government strictly constrained the parameters of daily life of the Appalachian "welfare mothers" whom Gilliom studied, its watchful and persistent eye leading to "hassle and degradation" that "hindered [the mothers'] ability to meet the needs of their families." These women were forced to adhere to restrictive rules while simultaneously finding it necessary to skirt those rules in order to survive. They thus experienced surveillance not as an invasive uncovering of their secrets but as a loss of "a great deal of the autonomy and control that they could otherwise have over their personal lives." As one explained, "All the time you are on welfare, yeah, you are in prison."[32]

The poor and minorities are also most susceptible to harms caused by a lack of privacy in dealings with the private sector. As companies increasingly make decisions about individuals based on data drawn from their online behavior and social networks, lower socioeconomic groups can be unfairly excluded from credit, jobs, housing, and healthcare in ways that circumvent anti-discrimination laws.[33] Low-wage workplaces monitor their employees' keystrokes, location, emails, and online browsing to detect unsanctioned behavior, which can result in firing.[34] The profiles created by data brokers (such as "suffering seniors" and "urban scramble") make it possible for companies to target those susceptible to predatory loans and scams with precision.[35]

Thus, as a 2018 book by Eubanks suggests, these privacy infringements and algorithms conjure up a new meaning for the acronym AI: not "artificial intelligence," but "automating inequality."[36]

* * *

The smart city represents the vast expansion of both government and corporate data collection. Embedding sensors, cameras, software, and an internet connection in everyday objects from streetlights to trashcans—creating what is known as the "Internet of Things"—makes it possible to collect remarkably precise data about what is happening in a city. This data could be used to facilitate beneficial outcomes: reducing traffic, improving infrastructure, and saving energy. But it also includes detailed information about the behavior of everyone within the city.

Smart city technologies make it easier than ever for municipalities to identify and track individuals. Sensors on streetlights and other forms of "smart" infrastructure (like LinkNYC kiosks) can track internet-connected devices in their vicinity, making it possible to follow people's movements throughout the city. Cameras paired with software that can identify people or objects generate additional surveillance threats. In Los Angeles, for example, automatic license plate readers (ALPRs) record the location of three million vehicles every week, collecting information that often finds its way into the hands of U.S. Immigration and Customs Enforcement

(ICE).[37] The push for police-worn body cameras, supported by many as a tool to hold police accountable, creates the potential for widespread surveillance by police of all public space: given that body camera manufacturers are developing facial recognition software to analyze this footage, and given that only one police department in the United States has a policy governing body cameras that "sharply limits the use of biometric technologies,"[38] it is likely that body cameras will soon be used by police wherever they go to track the movements of civilians, identify participants at protests, and scan crowds to see who has outstanding warrants.[39] Making similar use of technology, police in Orlando are using Amazon's facial recognition service to monitor in real time everyone who appears in the video feeds from traffic cameras.[40]

Meanwhile, for companies eager to expand their data collection beyond your browser and into physical space, the smart city is a dream come true. Many companies already possess the knowledge and influence necessary to restrict individual autonomy and exploit people, but if companies like Sidewalk Labs have their way, smart city technologies will vastly increase the scale and scope of data they collect. Companies that place cameras and MAC address–sniffing sensors on Wi-Fi kiosks, trashcans, and streetlights will gain heretofore unattainable insights about the behavior of individuals. And given the vast reach of hard-to-trace data brokers that gather and share data without the public's knowledge or consent, one company's data can easily end up in another's hands.[41]

Once these smart city technologies are installed, it is practically impossible for someone to avoid being tracked. Many defend online companies' mass data collection by pointing to the opportunity opt out: if you don't like data about you being collected, don't use the websites or apps that collect it. But since it is almost impossible to communicate, travel, or get hired without email, search engines, smartphones, and social media, this is an unreasonable choice. In the emergent smart city, with sensors and cameras on every street corner—remember, New York is slated for more than 7,500 LinkNYC kiosks—this argument reaches an even more perverse conclusion: if you want to avoid being tracked, you must opt out of public space.

This position places urban residents in an untenable bind. On the one hand, eschewing modern technology would mean not just forgoing public announcements and conversations that occur online but also losing out on services that governments distribute by analyzing data.[42] For example, if cities analyze people's movements with MAC address sensors to determine where to place bus stops, they will overlook the needs of those without smartphones (and those who turn off their phones to avoid being tracked). On the other hand, those with smartphones and other wireless technologies must suffer the consequences of being tracked; and in places where cameras are used to identify people, it is impossible to escape being tracked even by abandoning one's personal digital technology.

Such conditions would most significantly hurt the urban poor, who are already the most vulnerable to online tracking:[43] while well-off New Yorkers who do not want LinkNYC to track them can forgo free Wi-Fi in favor of a personal data plan, lower-class residents have no alternative to free Wi-Fi (indeed, the whole point of LinkNYC is to provide internet access to those who cannot afford to pay for it) and must accept being tracked in exchange for internet access. Thus, the inevitable outcome of accepting pervasive data collection in smart cities—and believing the myth that opting out is a legitimate option—is the creation of "a new type of societal class system: a higher class of citizens free from fear of manipulation or control and a lower class of citizens who must continue to give up their privacy in order to operate within the prevailing economic system, losing their ability to control their own destiny in the process."[44]

Smart cities thus provide welfare offices, police, employers, data brokers, and others who use data to control the lives of the urban poor with a new tool for surveillance and exploitation. A single mother could be flagged by an algorithm to lose welfare benefits because she was identified at a protest by body camera footage. A black teen could be identified for surveillance by the police because he connects to a public Wi-Fi beacon that is often used by people with criminal records. An elderly man could be targeted for predatory loans because his car was recently identified by automatic license plate readers as it was driven out of an impound lot.

The grave risks to equity, autonomy, and social justice introduced by data collection in smart cities raise new challenges for and impose new responsibilities on city governments. Beyond determining what data to collect for themselves, municipalities must also act as gatekeepers for private companies eager to access new environments for data collection. Many smart city projects are similar to LinkNYC and involve public–private partnerships through which municipalities procure technology from companies in order to offer new or improved services. For governments, working with companies makes it possible to take advantage of private-sector technologies that would be difficult to develop internally. For companies, partnerships with cities present a rare and incredibly valuable opportunity to place data-gathering sensors throughout public spaces. City governments must therefore thoughtfully consider whether the benefits of new services are worth the price of allowing companies to collect untold amounts of data about the public; if not, they must find ways to obtain the benefits of new technology without incurring those costs.

But even if cities collect data for a benevolent purpose or trust the private vendor whose technology they procure, they must grapple with the numerous ways in which sensitive information can be exposed to the public or to groups with malicious intentions. Once data is collected, it is liable to be released and abused. And even within government, as the example of Los Angeles ALPR data being shared with ICE indicates, data collected by one agency can end up in the hands of another. New technology that enables the collection of more granular and sensitive information magnifies these risks.

Over the past decade, many city governments have embraced "open data" initiatives, which involve releasing municipal datasets online in an effort to make government more transparent and accountable as well as to foster civic innovation. These efforts have led to thousands of datasets being released in cities across the country, paving the way for transit apps,[45] user-friendly tools to explore municipal budgets,[46] and countless hackathons. But because much of the data that municipalities collect relates to the people within those cities, open data also occasionally reveals sensitive information about individuals. By releasing open data, cities have inadvertently revealed the identities of sexual assault victims and people who carry large sums of cash at night,[47] as well as people's medical information and political affiliation.[48] Although there are strategies that cities can employ to reduce the risks of such disclosures, they must grapple with the inevitable tension between open data's utility (more detailed data provides greater transparency and can be used for more purposes) and risks (more detailed data contains more sensitive information), a dilemma that will only worsen as the scope of municipal data collection expands.[49]

Even when governments do not proactively release data, they often have few means to protect their information from becoming public knowledge. Federal and state public records laws, designed to enforce government transparency and accountability, compel governments to release data they control when requested to do so by a member of the public. Although these laws contain exemptions restricting the release of sensitive information, their reliance on the outdated PII and secrecy frameworks severely limits the scope of such exemptions. Thus, as cities gather and store supposedly anonymous data about people's behavior, they will be sitting on increasingly large piles of information that could easily be exposed to reveal sensitive information about people. Case in point: the NYC taxi trip data that was analyzed to infer the identities of strip club patrons was initially released through a public records request and then posted online by the requestor for anyone to use.[50]

Finally (and this is a concern for companies as well as governments), any data that is collected and stored can be released through hacks and security breaches. In 2017, cyber attackers stole the names, addresses, and credit card information of 40,000 residents of Oceanside, California, which they used to make unsanctioned online purchases.[51] The previous year, a hack of Uber exposed the personal information (including names, email addresses, and phone numbers) of 57 million users.[52] Moreover, the new sensors being installed in countless Internet of Things deployments to collect detailed data about urban conditions are egregiously insecure.[53] To the security technologist Bruce Schneier, cases like these dispel the prevailing narrative that all data is good and more is always better. Instead, he says, "[D]ata is a toxic asset and saving it is dangerous."[54]

The imperative to deploy new technology thus thrusts municipalities more strongly than ever into the role of being stewards of urban life. They must decide what data can be collected and who gets to access it (while also accounting for the fact that once data has been collected it may be exposed to others). Cities are therefore confronted not merely with technical judgments about how to operate municipal services but with deeply political decisions that will determine the future of urban life. Will cities increase their control over their inhabitants and provide similar control to companies without any public dialogue? Or will they ensure that the social contract they create through technology provides people with a right to the city free from being monitored and manipulated by corporate and governmental entities?

Through this lens, what is remarkable about LinkNYC is not that a Google-related company would provide a free service in exchange for collecting user data—monetizing such data is their fundamental business model, after all—but that the City of New York would allow them to, would sell off the public's privacy for what one local paper calls "chump change."[55] As the media theorist and author of *Throwing Rocks at the Google Bus* Douglas Rushkoff puts it, LinkNYC represents "a deal with the devil we really don't need."[56]

* * *

It is not just increased data collection that threatens to create an undemocratic social contract in smart cities. As we saw in the previous chapter, cities are increasingly using algorithms to inform core functions like policing and social services. In New York City, for example, algorithms are used to assign students to schools, evaluate teachers, detect Medicaid fraud, and prevent fires.[57] And despite the seemingly sophisticated nature of these algorithms, they are neither foolproof nor neutral: bias can arise both in the training data on which they rely and in how they are deployed.

Yet even though the decisions that algorithms inform are potentially life-altering, auditing their design and impacts is remarkably difficult. Chicago's Strategic Subjects List provides an instructive example: the Chicago Police Department has resisted countless calls to publicly share the details of how the algorithm works and what attributes of people it considers.[58] Police are therefore showing up unannounced at people's homes without ever explaining what led them there.[59]

Municipal implementation of algorithms thus raises grave concerns for urban democracy: cities typically provide the public with little or no insight into how their algorithms were developed or how they work. Cities rarely release the source code governing the algorithm or the data it learns from. The public may not even know when algorithms are being used.

In many cases, municipal algorithms are concealed because they are developed and owned by private companies with a financial interest in secrecy. Because public agencies typically lack the resources and technical sophistication necessary to develop algorithms on their own, they often contract with companies to procure algorithmic systems. And although there is value in relying on technical experts to develop algorithms, these new relationships shift decision-making power away from the public eye.

Through nondisclosure agreements and broad assertions of trade secrecy,[60] technology companies prevent the governments that utilize their services from revealing any information about those tools or their use. These companies include Intrado (which has developed Beware, software that police departments use to calculate people's "threat scores") and Northpointe (which has developed COMPAS, an algorithm that predicts one's likelihood to engage in future criminal activity, recently rebranded as equivant).[61] Even public records laws have been ineffective tools for shining light onto these proprietary algorithms. For example, two lawyers submitted public records requests for information about PredPol to eleven police departments reported to be using the software. Only three responded, and none provided substantive information about the algorithm or its development.[62]

As a result, governments may make consequential decisions about people without providing any transparency regarding how those decisions are made or any due process. Consider the case of Eric Loomis, who was arrested in 2013 in the city of La Crosse, Wisconsin, for driving a car that was involved in a shooting. Loomis pled guilty to fleeing the police, and the state used Northpointe's COMPAS algorithm to inform his sentencing process. When giving Loomis a six-year sentence, the judge explained, "The risk assessment tools that have been utilized suggest that you're extremely high risk to reoffend."[63] Because Northpointe claimed that its algorithm was a trade secret, Loomis was not permitted to assess how the algorithm made this prediction. His challenge of the judge's use of this opaque system was unsuccessful.

As cases like Loomis's become commonplace, Solove's reference to *The Trial* begins to appear prescient. Just as Kafka's Josef K. faced a trial in which he knew neither his crime nor his accuser, here was Loomis being given a sentence that was influenced by an algorithm that neither he nor the judge could inspect.

The deeper danger of such algorithmic decision making is that when governments use proprietary algorithms like COMPAS, the unelected and unaccountable developers of these systems are granted significant power to dictate municipal practices and priorities. We saw in chapter 4 how algorithms—and thus their effects—are shaped by judgments about what data to use, what input factors to include, and how to balance false positives and false negatives. These seemingly technical choices influence public policy; as governments increasingly make choices based on algorithms developed by private companies, they will increasingly make decisions based on the values and assumptions that those companies embed within the algorithms. For example, Northpointe's choice to predict people's likelihood to commit crimes in the future places criminal justice adjudication within the prosecutorial and racialized context of crime risk.[64]

One of the most important decisions that Northpointe made when developing COMPAS was how to ensure that its predictions were not racially biased. On even a purely technical level, making unbiased predictions is more complicated than it may first appear, as there are several technical criteria for "fairness" from which to choose. The company strove for what is known as "calibrated predictions," meaning that their model should be equally accurate for both black and white defendants. This is, on its face, a sensible choice. Yet in 2016, ProPublica revealed that COMPAS was twice as likely to misclassify a black person than a white person as "high risk," potentially leading to criminal sentences for black defendants that, without justification, are longer and more punitive.[65] This was seen as evidence that COMPAS was racially biased. Perhaps Northpointe should have optimized instead for "balanced classes," meaning that its algorithm would have equal false positive rates for black and white defendants. Doing so would have addressed the issue that ProPublica highlighted, but at the price of raising a new one in its stead: the new algorithm would no longer make calibrated predictions (i.e., it would be more accurate for one group than for another). In any attempt to make fair predictions about two groups for whom a phenomenon of interest—here, recidivism rates—occurs at unequal rates, this trade-off in unavoidable: it is impossible to attain both calibrated predictions and balanced classes.[66]

The point is not that COMPAS was wrong in prioritizing calibrated predictions: neither of its options for defining fairness was clearly superior to the other. Indeed, many policy decisions involve complex trade-offs of this sort. The problem is that this decision—which shapes a fundamental aspect of the criminal justice system, and hence people's lives, in every jurisdiction that adopts COMPAS—was made by the staff at Northpointe with no input from public officials or the broader populous.

Reliance on municipal algorithms thus represents a drastic change in how policies are developed and implemented. In the past, although such decisions were by no means entirely transparent and accountable, they were presumed to be political and to require democratic input, oversight, and justifications. Decisions made by computational systems (especially ones developed by private companies) eschew these obligations: it is in many cases impossible for the public to have any input into or wield any control over algorithmic decisions, if they know about the algorithms at all. And even when they do assert opinions, questions about how algorithms should be designed are often seen as technical matters best left to the "experts."

Compounding the dangers that emerge as governments use opaque and unaccountable algorithms is the vastly expanded data collection that smart city technologies enable. As previously unattainable data about individuals becomes available, much of that information may be incorporated into algorithms that influence criminal sentences and other significant decisions. In the court cases of the smart city, factors such as where you spend time, how late you stay out at night, and whether you participated in a particular protest—data that you may never have even known was being collected, and whose collection you certainly never consented to—could influence the sentence you are given.

As these manifold dangers come to light, one city has made a valiant attempt to alter how it deploys algorithms. In August 2017, James Vacca, a member of the New York City Council, introduced a bill that would require city agencies to release the source code of every algorithm they use to target social services, impose penalties, and inform the actions of police.[67]

Vacca, who had been involved in New York City government for almost four decades, was familiar with the public's lack of access to algorithms: his many attempts over the years to learn about the algorithms that dictate staffing in the police and fire departments had all been thwarted.[68] In a public hearing on his proposed legislation, Vacca explained his motivation for the bill. "I strongly believe the public has a right to know when decisions are made using algorithms, and they have a right to know how these decisions are made," he said. "For instance, when the Department of Education uses an algorithm to assign children to different high schools and a child is assigned to their sixth choice, they and their family have a right to know how that algorithm determined that their child would get their sixth choice. They should not merely be told that they were assigned to a school because an algorithm made the most efficient allocation of school seats. What is considered to be most efficient? Who decided this?"[69]

The final version of Vacca's bill, approved by the City Council in December 2017 and signed by Mayor de Blasio in January 2018, represents a curtailed enactment of his initial vision. The legislation established an "automated decision systems task force" to examine how city agencies use algorithms. The group will recommend procedures that would achieve several outcomes: increase public transparency about the algorithms being used, determine which algorithms should be subject to oversight, provide the public with opportunities to receive explanations about algorithmic decisions, and evaluate whether algorithms unfairly impact certain groups of people. The task force will present these recommendations to the mayor in a public report.[70]

Although the new law leaves much to be desired—the task force can merely make recommendations and has little power to compel public disclosure, especially when dealing with companies eager to protect their trade secrets[71]—it is a productive start, building momentum for policies that hold cities accountable for how they develop and deploy algorithms. Just as importantly, Vacca's efforts helped shift public discourse toward considering algorithms not as unassailable oracles but as socially constructed and fallible inputs to political decisions. Such a change is essential to the development of Smart Enough Cities.

* * *

When it comes to algorithms as well as data collection, municipal decisions must be grounded in democratic deliberation that provides the public with a meaningful voice to shape development, acquisition, and deployment. Such work will, perhaps counterintuitively, aid rather than hinder the adoption of technologies that improve life in Smart Enough Cities.

The Array of Things (AoT) in Chicago highlights the value of engaging the public to protect privacy. The product of a collaboration (launched in 2014) between the City of Chicago, the University of Chicago, and Argonne National Laboratory, the AoT is designed to be "a 'fitness tracker' for the city."[72] It will eventually consist of several hundred sensors installed throughout Chicago to track environmental conditions such as air quality, pedestrian and vehicle traffic, and temperature. In one neighborhood that has a highway running through it, for example, city officials hope to reduce asthma rates in children by using the data as a guide for where trees are most needed and where bus stops should be located.[73]

On the surface, the Array of Things appears quite similar to LinkNYC: it is a large-scale deployment of sensors that can collect vast quantities of data. The AoT could have led to public backlash if too few people trusted that the data was being collected and managed responsibly. Chicago took a vastly different approach from New York in deploying its sensor network, however: it designed the sensors to avoid collecting sensitive personal data while also directly engaging the public to share how it had done so and to collectively develop priorities. This is, in part, a benefit of how the Array of Things is owned and operated: whereas LinkNYC is managed by a private company and thus structured to maximize profit, the AoT is run by government and academic institutions and therefore focused on generating public benefits rather than revenue.

Chicago has made protecting public privacy an essential element in the AoT's architecture. While developing initial plans for deploying the AoT, the city convened a committee of local privacy and security experts to provide independent oversight regarding how the system collects and stores data. It then organized public meetings to explain to the broader Chicago population how the Array of Things worked, how the program protected privacy, and how the sensors could improve living conditions.[74] To fully incorporate public concerns about privacy into the policies that govern the AoT, Chicago also released a draft of its privacy policy for public comment. This draft garnered more than fifty inquiries, all of which the city publicly responded to and incorporated into the final privacy policy that governs the program.[75]

This outreach helped Chicago identify the public's privacy concerns and hold itself accountable to addressing them. For example, one major worry was that the sensor's cameras would collect images that could be used to track people's movements over time. Such tracking was not the city's intention—it sought to collect the images to obtain traffic counts (via analysis from computer vision software)—but it was certainly possible that images gathered from the cameras could be abused. Collecting that footage could violate privacy expectations and fuel residents' opposition to the entire AoT project.

In response, Chicago devised a thoughtful solution that draws on a practice known as "data minimization," which involves collecting and storing the minimal amount of information needed to achieve a project's goals. Data minimization can take several forms; two common tactics are ignoring extraneous features entirely (say, not collecting someone's location) and storing data in a deliberately imprecise format (recording someone's location simply as a zip code).[76] Because Chicago desired only the traffic counts that could be calculated from the images, there was no need to store the camera footage itself. The AoT sensors were altered to calculate this number, transmit only that value to the project servers for retention, and then immediately delete the images.[77]

Chicago's development of the AoT exemplifies how cities can synthesize public engagement and data minimization to make possible the deployment of cutting-edge technology in Smart Enough Cities. By allowing public input into how it implemented the Array of Things, Chicago ensured that the social conditions mediated by its new technology would be democratically determined and desired. When the public raised concerns, the AoT team found a way to collect the data that it required and no more, in such a way that the city could achieve its analytical goals without compromising the will or privacy of its constituents. Had it not followed these practices, the entire project might have been stymied by public opposition.

During the same years that Chicago was developing the Array of Things, Seattle was learning the hard way why it is necessary to protect privacy when deploying new technology. In November 2013, backed by $2.7 million from the Department of Homeland Security, the City of Seattle installed a network of sensors and cameras to monitor potential threats coming from its harbor. The public—who had received little warning about the new technology's installation or use—quickly became concerned when they noticed that the wireless sensors appeared capable of tracking individuals by recording the movements of their wireless devices. In response to questions about how the new sensors would be used, the Seattle Police Department replied that it was "not comfortable answering policy questions when we do not yet have a policy," which further inflamed tensions by suggesting that the city was being cavalier about individual privacy and not taking the necessary precautions to prevent undue surveillance.[78] As the public controversy mounted, even despite indications that the sensors could not in fact track people's devices, the police department shut down the program.[79]

The botched rollout "became a really good learning lesson," reflects Seattle Chief Technology Officer Michael Mattmiller.[80] In part because of a lack of familiarity with the technology and its risks, the city had deployed these new sensors without considering whether their architecture aligned with public priorities. "For those who aren't educated on how technologies work and potential privacy harms, it's very easy to just focus on the outcomes of a technology, not the means by which they achieve those outcomes,"

Mattmiller notes. Moreover, deploying technology with meager public outreach meant that the city had a poor understanding of how members of the public value their privacy. As existing privacy laws based on PII become obsolete, social conceptions of privacy evolve, and new technology makes it possible to collect ever-increasing amounts of data, cities have no reliable guide to tell them what level of data collection is appropriate. So as Seattle shut down a new technology program over public concerns about invasive data collection, the message became clear: without technological expertise, strong privacy policies, and public dialogue to address privacy risks, it would be almost impossible to improve municipal operations and urban life with technologies that involve collecting data.

Eager "to move forward on these missteps we've had with the public," Mattmiller created a Privacy Advisory Committee consisting of local technologists, lawyers, and community representatives. The group's task was to share public priorities and concerns related to data collection and to guide the city in developing practices that protect privacy. Through a series of public meetings, the committee developed six Privacy Principles—collectively affirming a commitment to transparency, accountability, and minimal data collection—intended to guide the city's collection and use of personal information.[81] The committee then helped the city develop a thorough privacy policy in 2015 that embodies these principles.[82] Through these efforts, says Mattmiller, the Privacy Advisory Committee "really ensured that the community's fingerprints and best practices were infused in our privacy program."

A key component of Seattle's privacy policy is a mandate to conduct a "Privacy Impact Assessment" every time it develops a new project that involves collecting personal data. The city must undertake a risk-benefit analysis that weighs the project's potential utility against its potential threat to individuals' privacy. The goal is to proactively highlight and mitigate expected risks without crippling the project, thereby enabling the city to balance its charge to enhance public welfare with its responsibility to protect civil liberties. These assessments help Seattle ensure that their projects adhere to the Privacy Principles; adjustments typically take the form of changing how data is gathered, stored, and shared to reduce the collection and exposure of sensitive information.

Mattmiller also emphasizes the importance of educating city staff about privacy risks and how to mitigate them. To help departments recognize

and prevent privacy harms, the city nominated a member of every municipal department to act as a "privacy champion." These individuals coordinate Privacy Impact Assessments and educate their colleagues about
best practices in following the city's Privacy Principles. When Mattmiller
and his team need to update city staff about recent developments or new
privacy risks, the privacy champions help disseminate this information
throughout departments. And in 2017, Seattle further institutionalized its
commitment to privacy by hiring a chief privacy officer, becoming one of
the first cities in the nation to have someone with a citywide mandate to
manage privacy.[83]

Adopting these practices meant that the next time Seattle prepared to
adopt a new technology that required collecting the public's data, it was
ready to do so thoughtfully and responsibly. The big test came when the
Seattle Department of Transportation (SDOT) deployed a thousand sensors to measure traffic flow and travel time between various locations. By
tracking the movement of wireless devices across town (via their MAC
addresses)—data that had never before been available to public officials—
SDOT hoped to spot patterns that would help it reduce congestion.

With their new privacy program in place, Seattle was able to delibera-
tively and openly weigh the costs and benefits of this technology. First,
Mattmiller and his team consulted with SDOT and the technology vendor
to ensure that excessive surveillance would not be the price of smoother
traffic. After identifying several privacy risks in the technology, they
pushed the company to implement data minimization approaches that
would make it harder to identify and track any individuals using the data.
And instead of working behind the scenes, as it had when deploying the
security network around the harbor, the city proactively shared what it
was doing and why. Mattmiller took the case for this technology to the
public, explaining, "If you like using Google Maps and you like seeing the
red, yellow, and green lines that show you traffic flow and reroute you
around congestion, then we need to collect data to inform those maps,
and we believe this is the least invasive way possible. If you don't like this,
here's a website you can go to to opt out, so if we see your cell phone we
ignore it."

By mitigating the most salient privacy risks and explaining the purpose
of its efforts, the city garnered public support rather than outrage, Mattmiller notes. "Very plainly conveying the value of what you're doing, being

transparent with privacy threats, and showing how you've mitigated those threats builds trust."

But it is not enough for cities to act in good faith—public oversight of municipal technology must be institutionalized. To further empower the public with control over how the city collects and uses data, Seattle enacted a surveillance oversight ordinance in 2017. The bill requires every department to hold public meetings and obtain City Council approval before acquiring any surveillance technology, to publicly describe how it uses surveillance technology, and to assess all surveillance technology for its impacts on privacy and equity.[84] The ordinance thus ensures that Seattle's decisions to acquire and deploy surveillance technology (whether hardware or software) are subject to robust deliberation by the public and elected officials rather than the shrouded decisions that are endemic across U.S. cities. In 2016, for instance, a local newspaper reported that the Seattle Police Department had been using social media monitoring software for two years without ever notifying the public.[85] Similarly, police in New Orleans used predictive policing algorithms for several years without having gone through any public procurement process, and even members of the City Council were left in the dark.[86] Joining Seattle in pushing back against this trend, dozens of cities across the United States (ranging from Hattiesburg, Mississippi, to Oakland) have passed or are developing similar surveillance ordinances,[87] building momentum for one of the most important tactics for transforming smart cities into Smart Enough Cities.

By recognizing that seemingly technical decisions about what data to collect are in fact political ones with massive implications for civil liberties and social justice, Chicago and Seattle both demonstrate how Smart Enough Cities can provide new services and enhance daily life with technology while at the same time fostering democratic social contracts. These achievements contradict the view through tech goggles, which falsely suggest a binary trade-off between privacy and innovation. According to this worldview, becoming smart means collecting and analyzing data to improve efficiency—and if protecting privacy and liberty requires gathering less data, then the smart city must be one without privacy or liberty.

In the Smart Enough City, however, where privacy is a human right essential to maintaining liberty and equity, protecting privacy enables rather than prohibits the use of new technology. "If you have a well-resourced, well-functioning privacy program, that program will promote innovation . . . and allow agencies to expand to new technologies," explains Marc Groman, senior advisor for privacy at the White House Office of Management and Budget under President Obama.[88]

While the smart city collects as much data as possible as it chases maximum efficiency, and the proverbial dumb city collects no data, the Smart Enough City collects data only after it has won public support to do so and after establishing privacy-protecting policies. The question for the Smart Enough City is not simply "What data should we collect?" or "How much data should we collect?" but "How can we accomplish our policy goals, with the aid of data, without violating public expectations or rights?"

*　*　*

In 2014, Nigel Jacob, co-founder and co-chair of the Mayor's Office of New Urban Mechanics in Boston, got a call from a local engineer. "I've been studying this whole parking problem, and it's actually very simple!" he excitedly told Jacob. The engineer had developed a clever solution to minimize the frustration of searching for on-street parking (and the congestion that this search creates): an app to pay for and reserve parking spots. Request a space before you leave the house, and a metal bollard would pop up and hold the space until you arrive. "It's just a question of resource allocation," the engineer explained.[89]

Whether or not this could work in practice—that is, make it easier to find parking—was beside the point, says Jacob. "We had this discussion where we explained that no one person has a right to public space. There's a social contract there. We started talking about that, and he started to see that it's a lot murkier." If increasing parking efficiency meant giving individuals the ability to reserve public space, Jacob was not interested.

The engineer in Jacob's story is typical of many technologists: they emphasize efficiency and convenience for some without considering the way in which those goals are achieved. Efficiency as an end seems to justify any means, or simply makes the means (and its by-products) irrelevant. And all too often, Jacob admits, cities buy into this logic and fail to consider the broader impacts. "We have a long track record of buying the wrong technology for a particular problem because we don't think about the politics of particular architectures," he laments.

When municipalities do not consider how the technology they acquire actually operates, the companies behind that technology dictate its architecture. As cities then deploy that technology, these design choices influence the social contract between people, companies, and the government. Under the guise of making a city "smart," companies sell technology that collects sensitive data and is opaque to the public—attributes that increase their profits—as if that were the only possible way that their products can function. Governments often then deploy that technology without sharing any details about it.

These are not inevitable outcomes dictated by the demands of new technology—they are instead the political arrangements desired by those who develop and control that technology. But as Chicago and Seattle demonstrate, alternative and more democratic architectures are available: it is possible to deploy pervasive sensors that improve urban life without collecting and abusing vast quantities of information about people. Similarly, the creation of an algorithm task force in New York and the enactment of surveillance oversight ordinances in Seattle and other cities provide a clear path for municipalities to reverse the trend toward becoming black-box cities.

Municipalities must accept the power and responsibility bequeathed to them as gatekeepers to information about the public and stewards of the public's privacy. Rather than merely embracing every new technology as manna from technological heaven, Smart Enough Cities are compelled by this new role to consider the risks of potential technological designs and reject the architectures motivated by police surveillance and extractive corporate practices.

When procuring technology from companies, municipal leaders must look beyond what a tool can accomplish and use their leverage to negotiate more democratic policies regarding privacy and transparency. After all, technology companies need cities more than cities need technology companies. Municipalities may gain some knowledge and efficiency from new tools and software, but we know by now that efficiency is not the most important ingredient for thriving urbanism; companies, on the other hand, need someone to buy their products. Given these dynamics, cities have the opportunity to assert themselves as market makers, acting both individually and collectively to shape the direction of smart city technology. Recognizing this power, in 2017 a coalition of 21 chief data officers published a set of guidelines for companies developing open data portals, and 50 mayors submitted a joint letter to the Federal Communications Commission in support of net neutrality.[90] Barcelona is also a notable pioneer in this regard: it has restructured contracts with several major technology vendors to enhance the public's ownership and control of data.[91]

If city governments do not take these actions, technology companies may continue to gain opaque and unaccountable private power over urban life. Already, companies such as Uber and Sidewalk Labs possess far more data than municipalities do about urban conditions, and companies such as Northpointe develop algorithms that inform highly consequential decisions. And as smart city companies accrue additional investment and profit, they will gain further leverage over cash-strapped municipalities: part of the reason smart city initiatives like LinkNYC are appealing to cities is that the latter lack the resources to provide public services themselves. If twenty-first-century urban residents are to have a right to the Smart Enough City with meaningful democratic control over technology production and use, municipalities must assert their authority over companies and be provided with the resources necessary to do so, while also themselves becoming more democratic.

Rushing to become a smart city may lead to new insights and efficiencies, but at the cost of creating cities in which the government and companies wield immense power over individuals by collecting data and making opaque decisions about them; poor and minority residents will be most subjugated. Yes, among the many responsibilities of cities are providing effective services and spending their funds judiciously. But to recklessly pursue technology that advances these goals without considering its full impacts is a severe dereliction of duty. For as we have seen, the benefits of new technology are often illusory and deploying it unthoughtfully can create more problems than it solves.

Cities can tread cautiously around technology without neglecting their responsibilities to care for the public, however. As we will see in the next chapter, technology can play a vital role in municipal innovations to improve urban welfare—but only when grounded in meaningful institutional and policy reforms that guide it toward the desired outcomes.

6 The Innovative City: The Relationship between Technical and Nontechnical Change in City Government

One of the smart city's most alluring features is its promise of innovation: it uses cutting-edge technology to transform municipal operations. Like efficiency, innovation possesses a nebulous appeal of being both neutral and optimal that is difficult to oppose. After all, who would want her city to stagnate rather than innovate?

Consider the homepage of Sidewalk Labs, which (as of October 2018) uses the word "innovation" five times. The company promises that it is "investing in innovation," will "accelerate urban innovation," provides "infrastructure that inspires innovation," and will "make Toronto [the site of its most ambitious project; see chapter 7] the global hub for urban innovation."[1] Elsewhere, the company has declared that "our mission is to accelerate the process of urban innovation."[2] Even more than technology, it appears that innovation is Sidewalk's key product. In this sense, innovation is of a piece with other smart city buzzwords like "optimization" and "efficiency": a vague but supposedly neutral and beneficial goal that is often touted by companies to advance their corporate agenda.

There is little doubt that cities could benefit from new ideas, policies, practices, and tools. Where smart city proponents like Sidewalk go astray, however, is in equating innovation with technology—or, to use Sidewalk's language, in concluding that "reimagining cities to improve quality of life" requires "digital advances to transform the urban environment."[3]

We will see in this chapter just how misguided that perspective is. It is wrong not only because technology alone cannot solve intractable social and political problems but also because of an attribute of city governments that we have observed but not yet fully explored: to derive benefit from technology, they must overcome institutional barriers by reforming policies and practices. This chapter will provide case studies of several cities—most notably, New York City, San Francisco, and Seattle—that demonstrate the painstaking processes required to improve governance and urban life with data. We will observe a very different relationship between technology and innovation than technophiles would ever recognize or praise.

* * *

In July 2015, public health officials in New York City identified an outbreak of Legionnaires' disease (an acute form of pneumonia) in the South Bronx. Seven people had already died and dozens more were infected. If not addressed immediately, the illness could spread throughout the Bronx and across New York City, threatening the well-being of millions.

The city's Department of Health and Mental Hygiene (DOHMH) quickly determined that the disease-inducing Legionella bacteria were incubating in the cooling towers that sit atop large buildings to support their air-conditioning systems. This is a common source for Legionnaires', especially during the summer when the use of air conditioning increases. As DOHMH cleaned the contaminated cooling towers, its staff recognized that a citywide inspection effort was necessary to prevent the disease from incubating in others. The City Council mandated that the city form a tactical response team to rapidly register and clean every cooling tower.

In many respects, this was nothing new for the most populous city in the United States. Led by NYC Emergency Management (NYCEM), city departments are adept at coordinating responses to crises ranging from hurricanes to terrorist attacks to citywide blackouts. But in the case of Legionnaires' disease, coordinating agencies was not enough—the city also had to coordinate multiple sources of data. Several critical questions loomed over the response effort: How many cooling towers are there in New York City? Where are they? Who owns them? Which ones are incubating Legionella? These questions could not be immediately or easily answered. Only a small fraction of buildings have cooling towers, and the city lacked any comprehensive database of their locations or owners.

So on a Friday afternoon a week into the emerging crisis, the Mayor's Office called Amen Ra Mashariki—New York City's chief analytics officer—asking for help.

"You can imagine this was an emergency of epic proportions," Mashariki says, looking back. "At the core of what we're supposed to do as government agencies is protect New Yorkers." Failing to quickly identify and inspect every cooling tower could allow Legionnaires' to spread out of control. Mashariki adds that what made this emergency particularly daunting is that addressing it required "a dataset that no one has ever considered. No one in City Hall or the Department of Buildings wakes up in the morning saying, 'We need to make sure our cooling tower dataset is primo because there may be an emergency that involves cooling towers.' This is a dataset that virtually didn't exist, and we had to cull it together."[4]

Luckily for New Yorkers, Mashariki's unique personal and professional background had prepared him for this moment. Growing up in a middle-class family in Brooklyn, Mashariki was strongly influenced by both of his parents. His father was a Vietnam War veteran and a social activist who founded a nonprofit to assist other veterans. His mother was a human resources executive at IBM, a position that afforded Mashariki access to some of the first PCs ever made. As a child, he was obsessed with computers and video games—he could not wait to learn how to program Donkey Kong—and his mother made a habit of bringing him to the office during school vacations. She taught herself how to code in BASIC (an early programming language) and began teaching her son when he was in fourth grade.

After studying computer science in college, Mashariki took a highly coveted job at Motorola in Chicago. He was building a successful career there, developing security protocols for two-way radios, when the Twin Towers were hit on September 11, 2001. The next day, when work resumed and the office was functioning as usual, Mashariki began to question how his work affected the world. "If something happens that changes the world but my job doesn't change, then the corollary must be true, which is my job doesn't have an impact on the world," he concluded.

Mashariki may have been developing technology that would help lead to the smartphones we all rely on today, but developing cutting-edge technology was no longer fulfilling for him. Following the lead of his activist father, Mashariki decided on that day that "anything I do from here on out has to explicitly have impact."

After spending most of the next decade in medicine, creating software for surgical robots and analyzing cancer treatment data, Mashariki plunged into government in 2012 as a White House Fellow in the Office of Personnel Management (OPM). The first ever computer scientist to be a White House Fellow, Mashariki entered with swashbuckling confidence that his technical expertise would help solve all the government's problems. "When I came in, I was like, 'I'm going to be the hottest thing,'" he recalls, cringing at the memory of one speech in which he announced his intention to use algorithms "'to fix how you guys do things and blow up the way you're thinking about problems.' I thought for sure I was going to come in here and be a superhero," he says. "And I remember looking around like, 'Why aren't they really digging this?'"

Mashariki entered government, in other words, like a typical technologist: confident that cutting-edge technology was the solution to many of government's challenges and that providing technical expertise would make him a savior. But his early efforts at OPM floundered because the solutions and approaches he espoused were poorly suited to the agency's needs. He was too focused on wielding technology in every situation rather than understanding the problem.

"Needless to say, I got my ass handed to me so many times, in many different ways," Mashariki recalls with a laugh. Whenever he suggested a technology that he thought would provide a quick and obvious fix, Mashariki was shot down because his colleagues had already considered that technology and determined that it would not address their needs.

These experiences helped Mashariki remove his tech goggles and realize that solving government issues with technology was much more difficult and complex than it had initially appeared. He realized that the issues he had diagnosed as technology problems were actually related to organizational capacities and needs, and that the key to addressing them was working with people and institutions rather than building technology. Mashariki also saw how bureaucracy, so often maligned as the force that stops innovation, prevented the implementation of bad ideas. Contrary to his expectations, working within the system was more productive than blowing it up. Mashariki's preconceived skepticism of government faded, leaving him with "a high level of respect for public servants."

Mashariki was named OPM's chief technology officer in 2013 and was handed responsibility for the massive project of digitizing the federal government's retirement process. Whereas a year earlier Mashariki would have concentrated on identifying the best software and persuading his colleagues to adopt it, now he recognized that success required bringing people together and focusing on institutional needs. He lists the many factors that he needed to consider: "You have to build relationships. You have to build consensus. You have to identify the people that you have to influence. You have to identify the people that you have to get insight from." Mashariki also knew not to dismiss the expertise of others within the organization. In the face of extensive doubts about the project from his colleagues, he built trust throughout OPM by emphasizing, "We're not here to tell you how to do your job. We're here to help you, learn how you do your job, and provide some capability for you." Mashariki's people-first approach was highly successful: over six months, his team achieved more progress than others had in OPM in the previous fifteen years.

Mashariki left OPM in 2014 to become New York City's chief analytics officer and director of the Mayor's Office of Data Analytics (MODA), the fledgling municipal analytics shop that the city had established the previous year. Having learned at OPM about the limits and potential of technology to improve government, Mashariki was eager for "the challenge of being the data guy for one of the largest cities in the world. Who wouldn't want to take that on?" He knew it would be the most demanding role of his life, but, Mashariki says now, "I had no idea just how complex it was really going to be."

Mashariki was only nine months into the job when Legionnaires' disease broke out. The scale of the task and the precision required to complete it were overwhelming. New York has more than a million buildings; given the city's limited human and financial resources, visiting each one to check for a cooling tower would take years, allowing the bacteria to fester and spread. But the city had to be comprehensive in its search. "This can't be 98 percent confidence that we got all the buildings," Mashariki explains. "This has to be 100 percent confidence." Mashariki's job was to accelerate the pace of inspections by using data and analytics to identify the buildings most likely to contain a cooling tower, and thus the buildings on which the inspection and cleaning teams should focus.

Unfortunately, synthesizing all of the city's data to form a coherent picture proved more difficult than anyone had predicted. For instance, it initially appeared that the Department of Finance (DOF) possessed the necessary data, since it tracked some buildings with cooling towers as receiving a tax write-off. But this dataset did not include every cooling tower, nor did it contain the names and contact information of building owners— information that was needed to verify the presence of a cooling tower, register it, and inspect it. And while the Department of Buildings (DOB) does collect information about building owners, it was incredibly difficult to align the two datasets, because DOB identifies buildings by address whereas DOF does so by tax parcel. In addition, DOB recorded the number of buildings that use a cooling tower, overlooking that some cooling towers service multiple buildings and some buildings are serviced by multiple cooling towers. MODA's first responsibility was to synthesize these conflicting and incomplete datasets, but despite their painstaking work the team could piece together only an incomplete list of cooling towers and their owners.

These gaps and disparities in data are common in city governments: although many departments collect nominally related data, each typically interprets and documents that information differently. Datasets collected by different departments are rarely designed to be synthesized. Every administrative division has its own IT systems and data structures, which are tailored to its individual needs and missions. This facilitates everyday tasks but hinders efforts that require merging data from multiple departments.

"A lot of people don't realize that there are different ways to count entities in the city," Mashariki explains. "Oftentimes you think you're counting the same thing but these two agencies are counting different things and they report it to the leadership of the city two different ways. If you don't have a team like MODA there, then it can be mayhem."

There was no room for such mistakes in this crisis, which required many sources of data. DOB created a website where building owners could register their cooling towers. The city's 311 call center contacted building owners to ask if they had a cooling tower. NYC Emergency Management canvassed the city as part of a public awareness campaign. Firefighters traversed the city, inspecting buildings to see if they had a cooling tower. The Department of Health and Mental Hygiene tested and cleaned the cooling towers that were identified.

The Mayor's Office of Data Analytics became the glue holding these rapid-fire efforts together. Every morning at 7 a.m., MODA would tell each agency where its resources for outreach or inspections were most needed that day. Departments would spend the day working on these tasks, recording data along the way. By 11 p.m., MODA would receive reports about each agency's progress—at which point it would assess the response effort's progress and determine each agency's tasks for the next day. Mashariki and his team became accustomed to sleepless nights.

MODA's next step was to synthesize these disparate and imperfect streams of information to quickly yet accurately identify every cooling tower in New York City. Early in the crisis response effort, only 10 percent of buildings visited by the city's inspection and cleaning teams had cooling towers—the outreach effort was wasting an enormous amount of time. Given such a low hit rate and the more than one million buildings in New York City, it could take years to find every cooling tower. To speed up the effort, MODA began developing machine learning algorithms that identified which buildings were most likely to have cooling towers by comparing their characteristics to those of buildings already identified as having cooling towers.

Despite the advanced data analysis required, MODA could not succeed by treating this as a purely technical challenge. Fortunately, Mashariki and his team were collaborating with other municipal agencies rather than focusing only on optimizing their algorithm. MODA's first list of potential cooling tower locations contained 70,000 buildings—a good start, but still too many buildings to inspect if they hoped to win the race to prevent more people from becoming ill, or worse. While reviewing this list, however, a few firefighters picked up on a key detail that the analytics had missed: the local fire code prohibited cooling towers on buildings with fewer than seven stories. When MODA incorporated this information into its algorithm, the list of potential cooling tower locations was cut in half.

"Your machine learning algorithm would not know stuff like that," Mashariki explains. "We would have been probably futzing around with a larger dataset if those folks didn't say, 'No, you don't have to go to those buildings.'" Whereas his younger self would have expected to save the day with a sophisticated algorithm, by this point in his career Mashariki understood that data and technology cannot solve every problem on their own. So even at a time when the city needed accurate and precise data to save lives, he reached out beyond the realm of databases and analytics to access as much contextual knowledge as possible. "You come in with your fancy machine learning algorithm in your pocket," Mashariki observes, "but what's always going to be your ace in the hole is the knowledge of the people who actually do the real work."

With the aid of contextual knowledge from other agencies, MODA's machine learning algorithm identified cooling towers with 80 percent accuracy—eight times the hit rate achieved by the city before incorporating analytics. The algorithm provided the city with the guide that it needed to identify, inspect, and clean every cooling tower in NYC within several weeks, stopping the outbreak by mid-August. The toll was significant—with 138 illnesses and 16 fatalities, this was the largest Legionnaires' outbreak in New York City's history[5]—but it would have been far worse without the efficient response effort that MODA made possible.

* * *

The Legionnaires' outbreak was a "game changer," according to Mashariki. The challenges experienced during the response effort highlighted major gaps in data quality and utility that could paralyze NYC in future crises. The next time an emergency arose—and Mashariki knew that there would be a next time—the city would need to respond more effectively and efficiently. The fire department might not have the spare capacity to traverse the city collecting data. A day spent reconciling discontinuities between datasets could slow the response effort and allow a crisis to intensify.

Because Mashariki knew that it would be impossible to precisely predict what the next emergency would be and what information would be essential—factors that he calls the "unknown unknowns"—he realized that the city needed to do more than just clean up a particular dataset or collect a specific new type of information. Instead, municipal departments across New York City would have to improve their data infrastructure and to develop generalized data skills so that they could better access, interpret, and use data for any purpose.

Mashariki adapted a page out of the city's existing playbook. One of NYC Emergency Management's responsibilities is to conduct emergency drills (akin to fire drills, but for municipal crises), in which multiple agencies practice responding to emergencies such as heat waves, coastal storms, and blizzards. These exercises are low-risk opportunities to identify gaps in services and coordination so that city agencies are prepared to act and work together when real crises occur. Following NYCEM's lead, Mashariki developed a similar training mechanism, called "data drills," during which departments could practice sharing data and using analytics to support the municipal response to an emergency.

The first data drill, in June 2016, brought together a dozen agencies to address a hypothetical blackout in Brooklyn. Every elevator in the area was shut down, leaving people stuck inside buildings and in need of rescue. City departments were required to synthesize data from multiple agencies to determine the location of every elevator in the region, predict which buildings had a population that was likely to be injured in those elevators, and develop a dispatch strategy to quickly send emergency response vehicles to those locations. The next drill, a few months later, involved the aftermath of a coastal storm and prompted agencies to assess the damage by integrating new data from post-emergency inspections with existing databases. The third data drill emphasized data sharing, enabling municipal agencies to practice how they access and use data from different departments during fast-paced crises.[6]

These drills are essential because, as MODA sees it, departments will be unable to improve operations and life in New York using data until they can manage and understand the pertinent information. By creating opportunities to work with data across a variety of situations, data drills push NYC municipal staff toward more effective and impactful uses of data. Departments have learned what information is collected by other agencies and how to prepare their own data so that other agencies can use it. To further aid these efforts, MODA is developing technical tools that make data easier to interpret and access. The team's first major project is a comprehensive Building Intelligence tool kit that unifies seven agencies' data about buildings into one interactive system, freeing departments from the burden of having to painstakingly make sense of conflicting information about buildings from different agencies. Data drills have also helped departments become more skilled in analyzing and applying data, whether to handle emergencies or to improve daily operations. And as these practices, processes, and tools permeate City Hall, they enable MODA to help departments serve New Yorkers more effectively. In one project, for instance, the team used machine learning to help the Department of Housing Preservation and Development proactively prevent landlords from harassing and forcing tenants out of rent-controlled apartments.

Mashariki's data drills exemplify how a city can become Smart Enough and illustrate the benefits of doing so. It is impossible to predict precisely what data and algorithms will be needed in the future—cities are too complex. But we can certainly predict the types of problems that will arise and the challenges that will accompany the use of data: poorly managed datasets that are inaccurate or incomplete, a lack of data fluency throughout

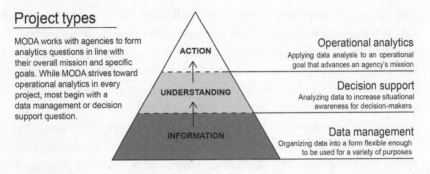

Project types

MODA works with agencies to form analytics questions in line with their overall mission and specific goals. While MODA strives toward operational analytics in every project, most begin with a data management or decision support question.

ACTION

UNDERSTANDING

INFORMATION

Operational analytics
Applying data analysis to an operational goal that advances an agency's mission

Decision support
Analyzing data to increase situational awareness for decision-makers

Data management
Organizing data into a form flexible enough to be used for a variety of purposes

Figure 6.1
The project pyramid that New York City's Mayor's Office of Data Analytics uses to guide its strategy.
Source: NYC Analytics, "Mayor's Office of Data Analytics (MODA)," p. 1, http://www1
.nyc.gov/assets/analytics/downloads/pdf/MODA-project-process.pdf.

departments, and the inability to synthesize information across datasets. These are not at their core technology problems, but to use new technology effectively they must be addressed.

As we similarly observed in Johnson County (see chapter 4), these issues typically arise because municipal departments and agencies operate as largely independent entities: each collects data appropriate for its particular tasks and responsibilities, without considering what data another department has. Two agencies might monitor the same aspect of the city but record information in a way that makes it difficult to match records. And because data has not traditionally been seen as a resource for analysis above and beyond its immediate purpose, there has been little reason historically to enforce data quality standards or uncover datasets that are tucked away on each department's computers. Moreover, municipal staff, who typically lack training in data analysis, are often wary of external attempts to improve or alter their work with technology.

In San Francisco, Chief Data Officer Joy Bonaguro is charged with overcoming these obstacles to make data a more valuable resource. With little patience for buzzwords and a no-frills attitude, Bonaguro effectively resists the allure of tech goggles. This resistance is necessary because, although her ultimate goal is to help the City of San Francisco use technology more effectively, the challenges Bonaguro faces are related primarily to people and policy.

Bonaguro's background in design makes her particularly attentive to the needs of technology users rather than to the capabilities of technology. And as a self-described *"Harvard Business Review* junkie," Bonaguro is comfortable dealing with complex bureaucracies such as city government. These perspectives help Bonaguro focus on using data to improve her city without getting caught up in the hype. "Smart cities are very technology-centered and technology-driven, and that's almost never a good strategy," Bonaguro says. "The reason that we're doing data science is not so we can be cool. We want to demonstrate that this is a tool we should be using."[7]

Since becoming chief data officer in 2014, Bonaguro—along with her team, DataSF—has had the mission of systematizing effective data infrastructure and governance throughout City Hall. They began by asking every department to create and share a data inventory, requiring them to catalog every data source and dataset they managed. By March 2015, thirty-six out of fifty-two departments had completed a full inventory; as of October 2018, 916 datasets have been cataloged.[8]

DataSF's next step was to make these datasets more accessible across departmental boundaries. In many cases, they could be released publicly on San Francisco's open data portal—enabling any department (as well as any member of the public) to access them without needing to go through bureaucratic channels or arrange data-sharing agreements; more than half of the inventoried datasets have been published as open data. For more sensitive datasets that cannot be released to the public, data-sharing agreements have been developed as necessary. But that process can be arduous: one such agreement, which will enable all the local health and human service agencies to coordinate delivery of their services, took more than a year to establish.[9]

The next phase of DataSF's ongoing efforts was ensuring that the city's data is of a high enough quality to facilitate analytics. This means that datasets should be accurate, up-to-date, consistent, and complete—a dataset about, say, cooling towers that omits records or has not been updated for several years would be of little use. But because city staff rarely employ their administrative data for analysis, they are not trained to consider these attributes. Bonaguro is therefore educating them in the tenets of data quality and how to achieve it. In 2017, DataSF released a guide titled "How to Ensure Quality Data," with an accompanying worksheet that walks staff through the steps necessary to assess and improve data quality.[10] The team is also training departments on data profiling, a technique for evaluating the integrity and reliability of datasets. Showing departments the limitations of their data has quickly paid dividends, Bonaguro says. "We did a test where we profiled a department's data and we brought it to a meeting. Their eyes bugged out. They'd never looked at their data that way." Shocked by the poor quality of their data, staff in that department have been following DataSF's new quality guide. DataSF has also created citywide standards for commonly occurring fields such as dates and locations to make it easier to match records and aggregate statistics across datasets.[11]

Creating an ecosystem of well-curated, accessible, and high-quality data required years of heavy lifting throughout San Francisco City Hall. Yet this was just the foundation of a larger effort: the data adds value only when it actually helps departments provide better services and governance. With that in mind, Bonaguro has recently shifted her focus toward training departments to improve their operations using data.

It is in this effort that Bonaguro's background in management and user design—together with her humble personality—has proved most essential. If DataSF were to approach departments claiming that data has all the answers, she says, "You wouldn't even be laughed at. You'd just be ignored." Departments would simply choose to avoid working with DataSF. "You really need to focus on developing relationships," Bonaguro adds. "These people know their business and so we have a lot to learn from them."

Bonaguro also knows not to immediately push departments toward the most sophisticated uses of data. After all, making decisions in a new way can require significant operational changes. And if staff are not familiar with data and algorithms, some will be threatened or insulted by suggestions that these technologies could improve their work, in part as a response to prior bad experiences with technologists who came in with little respect for the practices and expertise of existing staff. Recognizing these barriers, Bonaguro strives to learn about departmental needs and to "meet people where they're at." In partnership with the city's Controller's Office, she created a program called Data Academy to "lift data skills and capacity across the city," with courses that teach skills such as using databases, visualizing data, and creating information dashboards. "We think about everything in terms of gateway drugs. Data Academy is a gateway drug," Bonaguro says. "It's this continuous story of moving people up the 'data-use chain.'"

These tactics have helped Bonaguro develop partnerships with almost every department across the city. While some are eager to use data, others are resistant to disruption and outside influence. To showcase the value of data and demonstrate her good-faith intentions, Bonaguro starts with small projects that address each department's priorities and needs. She asks questions such as "What are the key questions you feel like you can't answer easily, or that you're answering over and over again?" and uses the responses to create dashboards that track and visualize metrics of interest—thus demonstrating how data could improve operations by breaking departments out of inefficient reporting practices that left data difficult to access and interpret.

Bonaguro recalls one department that spent most of their first meeting yelling at her. But she remained attentive to its needs and worked with staff to develop several dashboards that helped them monitor their performance. The department quickly warmed up to Bonaguro and has since made great strides in performance through better use of data. "That's how you move them to the next level," she explains. "You've solved something so you have a basis of trust on which to build the next step. How do you discover that? Through user research and design thinking, not through technology thinking."

Even once departments have recognized data as a valuable resource, much work remains to be done. We have seen several times that choosing what metrics to monitor and optimize for is a difficult and consequential task. Many well-intentioned efforts to use data in government go wrong because they fail to synthesize the glut of available data into metrics that appropriately capture their goals. "Metrics are tricky. Most metrics are bad," Bonaguro says, adding that when you choose the wrong metric, "then you're working toward the wrong thing."

All too frequently, Bonaguro notes, departments track metrics related to the quantities and processes behind their operations yet overlook the actual impacts of those operations and the desired outcomes. Instead of asking departments how many people they served last year, Bonaguro asks, "Did you serve them well? What happened as a result?" For as the next section of this chapter will describe, social service agencies that focus on how many people they serve rather than the impacts of those services will flounder. DataSF is therefore creating a Data Academy course to help departments design metrics that are tailored for their specific operations and goals. The course divides metrics into three categories: how much was done (quantity), how well it was done (quality), and who is better off as a result (impact).

Bonaguro sees all of this work as creating "fertile ground" for the most sophisticated stage of using data in city government: applying machine learning to improve operations. DataSF launched a program in 2017 to help departments use data science, and several quickly got on board.[12] The Department of Public Health built a predictive model to identify mothers who will drop out of WIC (Women, Infants, and Children, a federal program that provides services to low-income pregnant women, recent mothers, and young children) so that it can identify program barriers and make changes to better aid the women and their children.[13] In another project, the Mayor's Office of Housing and Community Development created an algorithm to flag eviction notices that appear anomalous or unlawful so that the city's eviction prevention services can intervene and keep residents in their homes.[14]

Bonaguro's work in San Francisco closely mirrors Mashariki's efforts in New York City. What makes them both exemplary Smart Enough City leaders are not their technical skills but their ability to pair technical acumen with a firm grasp of municipal needs and operations. For although enthusiasts of smart cities typically focus on the value that machine learning algorithms can unleash, such benefits cannot be realized without a long and arduous process of governance and institutional change: creating data inventories, bridging gaps between departments, and training staff to manage and use data.

Even then, insights from data cannot be translated into social impact without the traditional government operations that are so often maligned by technophiles. In New York City, for example, the Mayor's Office of Data Analytics provided invaluable information and analysis that aided the response to the Legionnaires' outbreak—but analytics did not, on its own, resolve the crisis: NYC Emergency Management coordinated the activities of several agencies, the fire department inspected buildings for cooling towers, and the Department of Health and Mental Hygiene tested and cleaned cooling towers. These activities were essential to curbing the spread of Legionnaires' disease. MODA helped direct these efforts, but success in preventing further illness ultimately depended on the work and expertise of other agencies.

"I don't want to make it come off as if MODA was this superstar," says Mashariki after sharing the story about the Legionnaires' outbreak.[15] "Data and analytics don't solve the problem: they support and add value to the people in your city that do solve the problem. Finding cooling towers in New York was like looking for needles in a haystack. MODA's job wasn't to find the needle—our job was to burn down the haystack to make it easier for the people who are actually doing the job to find the needle."[16]

* * *

Teams such as MODA and DataSF are crucial because poor data management and a lack of coordination across agencies can doom even the most well-intentioned efforts. That is exactly what happened in Seattle, where the Human Services Department (HSD) identified major gaps between its attempts to curb homelessness and the outcomes it was generating.

In 2015, Seattle's homeless population surpassed 10,000 people. Almost 4,000 of that group were living unsheltered on the street—a 38 percent increase from 2013, and the fourth straight year in which Seattle's unsheltered population had grown.[17] Dozens of homeless people died every year, and thousands of children were homeless.[18] As the city concluded a ten-year initiative (commenced in 2005) to end homelessness, it was clear that the situation facing Seattle was "worse than ever."[19] Local leaders declared a state of emergency.

For homeless mothers in Seattle like Shakira Boldin, accessing services for herself and her son was a constant struggle. "I would call programs and they would either be full or wouldn't have space or wouldn't be able to take me and my young child," she recalls. Local service providers lacked the resources and coordination to give Boldin's family the support it needed to stay safe and escape homelessness. "I had to have my son in a really volatile environment," she says. "We were sleeping on mats on the floor, and I didn't have anywhere to go."[20]

The HSD, which plays a primary role in maintaining the local social safety net, knew that drastic changes were necessary. Although the City of Seattle does not directly run any homeless shelters or other programs, it provides funds to community-based organizations—known in this context as "service providers"—to operate services such as shelters, hygiene centers, and meal programs. HSD was spending $55 million each year to fund homeless services, and yet families like Boldin's were falling through the cracks. Hoping to determine how it was performing and where it was falling short, HSD undertook a detailed analysis of its investments in homeless services.[21]

"We needed to do a thorough investigation to see how those investments were performing," says HSD Deputy Director Jason Johnson. "What we found was we couldn't always tell. We did not always have the level of information that told us whether a program was successfully moving people out of homelessness and into permanent housing. That's where the 'Aha!' happened," Johnson says. "We weren't able to tell the full story of the impact that these investments and programs were having on individuals. That's what we needed to do."[22]

In discovering that it could not even ascertain how its efforts actually affected people's lives and which of its programs were providing effective services, the Human Services Department realized that data about its homelessness programs was woefully incomplete.[23] Information was split across three separate data systems, forcing redundant data entry and preventing a cohesive portrait of services and their impacts from emerging. And because HSD did not clearly articulate the need for or value of data, service providers reported incomplete or unreliable information. This further reduced HSD's interest in the data, creating a downward spiral that left providers feeling justified in their poor reporting practices.

Such meager data management made it difficult for HSD to answer even simple questions about homelessness. Determining how many people had been served a meal required a manager to coordinate with ten program specialists and manually add numbers from separate spreadsheets. Answering more consequential questions, such as which families were in permanent housing after receiving services, was impossible. "On both the funder side and the provider side, there was way too much energy spent simply on collecting and reporting numbers and data," Johnson recalls.

This is not to say that HSD entirely lacked data about homeless services. "We had a lot of data" from providers, Johnson explains, but it was mostly "a bunch of 'widget counts' like how many people they've served and the demographics of those people."

"Seattle had wound up with this patchwork quilt of outcomes," says Christina Grover-Roybal, a Fellow from Harvard University's Government Performance Lab who helped HSD assess and overhaul its homeless services.[24] The city evaluated programs unsystematically, using a mishmash of criteria: showers taken, number of people who received services, amount of food distributed, exits to permanent housing, and so on. Even services of the same type (say, emergency shelters throughout the city) were often assessed on distinct metrics. Thus, Grover-Roybal explains, HSD "really couldn't compare performance across programs even if they were in the same service delivery model."

Service providers also struggled under this system, Grover-Roybal adds. "Sometimes one provider would have the same programs but run two different shelters, and those two different shelters would have different outcomes they were being held accountable to by the city. So as a service provider, they don't know what they're trying to achieve. We needed to get Seattle to monitor consistent outcomes across every single homeless service delivery program."

Because Seattle had not specified to service providers what it wanted to achieve, every service provider was working toward different goals. "We were not always clear upfront about the outcome that we were trying to achieve," reflects Johnson. "We were operating on the assumption that everyone was trying to get individuals and families into permanent housing, but in practice that was not always true. Services were helping people manage and mitigate the survival risks, but not necessarily trying to actively end their homelessness by getting them into housing."

Remedying this situation required Seattle to reform how it structures and manages contracts with local service providers. "Honestly, I feel like contracting was the only tool we had," Johnson says. "There were no other changes that we could implement to change service delivery, how data was collected, or how we looked at performance. Our only tool to do that were the contracts."

Government contracts are granted through a process called procurement: when Seattle decides that it wants to provide a social service, it requests proposals (or "bids") from companies and nonprofits. The city reviews these proposals and chooses to work with the organization that submits the best one (typically defined, whether explicitly or not, as the lowest-price proposal). Seattle then signs a contract with the winning bidder, after which it provides funding to that organization in exchange for that group's providing the desired program or service.

Seattle's dependence on contracts is not unique: governments across the United States rely on contracts to complete many of their most essential tasks, explains Laura Melle, the senior procurement lead for Boston's Department of Innovation and Technology. "Contracts are an input to every single output," she says. "A lot of people don't realize that government doesn't deliver our core services from scratch"; instead, whether paving streets or designing a website, "we're actually doing that in partnership with a private-sector company. Our role is often to select partners and manage contracts, with the goods and services provided by the private sector."[25] One estimate suggests that on average half of a city's budget goes to procured goods and services.[26]

Contracts, in other words, are the tools that turn the government's policy vision into reality. "A lot of really smart people have a lot of great ideas, but how do we make it happen?" Melle asks. "Whatever the great idea is, contracts are how you translate that into something that actually works for people the way that it was intended to." Effective contracts can bolster valuable government programs, while poorly constructed or managed contracts can doom even the best-designed policies.

Unfortunately, because procurement and contracts are typically seen as administrative and boring, the latter outcome is far more common.[27] Procurement processes are highly regulated and rigid, unattractive features that sharply reduce the quality and quantity of proposals that governments receive. And instead of being structured to incentivize desired performance outcomes, government contracts are typically managed with an emphasis on affordability and basic compliance.

Contracts had long been poorly managed in Seattle. When the Human Services Department reviewed its homeless services contracts, it found that a vast tangle of poorly defined goals and disconnected programs had accumulated over the years—more than 200 contracts with sixty homeless service providers.[28] Every time in the past that HSD had wanted to expand or provide new services, it received a pot of money from City Council and signed a new contract with a local service provider for that specific purpose. These contracts were never thereafter reprocured or restructured; the resulting jumble of contracts made providing and evaluating services difficult. "Service providers that had been around long enough had lots of different contracts for lots of different pieces, based entirely on how City Council had been divvying up money for the last ten to fifteen years," explains Grover-Roybal. Some contracts were more than a decade old. Some providers had numerous contracts for closely related services.

This "administrative nightmare," as Grover-Roybal calls it, made it difficult for service providers to effectively meet people's needs. "Even though
the service providers didn't necessarily think of all of these programs as
separate," she explains, they had to rigidly allocate staff time and other
resources according to their specific contracts, regardless of what services
would actually have the most impact. Moreover, because contracts were
never adjusted after their initial procurement, providers could not adapt
their services to meet the community's changing needs. This created a situation, Grover-Roybal says, in which "there are some shelters that are frequently underutilized and some shelters that are frequently overutilized.
But the way that it's set up right now, each shelter is limited to the size it
was when HSD did the initial request for proposals, and that could have
been five to ten years ago."

Johnson points to the local Young Women's Christian Association
(YWCA) "as emblematic of this issue." Over the years, the YWCA had collected nineteen separate contracts for addressing homelessness. Managing
these contracts took three dedicated staff members at the YWCA and four
more at the city. More importantly, the artificial barriers raised by these
contracts prevented the YWCA from most effectively serving those in need.
Shakira Boldin and her son might walk in and be the perfect fit for a program designed for families—but if the YWCA had already spent the money
allocated to that program, it could not pull unspent funds from another
program because the two were governed under separate contracts. Boldin
and her son would be left without help.

To counteract this problem, the Human Services Department developed
a novel approach: "portfolio contracts" that consolidate their previously
separate contracts with service providers so that funding could be allocated
more flexibly. Instead of having distinct contracts for each of their programs, providers now have a single contract with one pool of funding to
cover a larger portfolio of services; the first pilot of this approach merged
twenty-six contracts (totaling $8.5 million per year) into just eight.[29] This
is the "biggest win," according to Johnson, because it "allows agencies the
flexibility to move city money to where the individual that they're trying
to serve needs that money to be."

The introduction of portfolio contracts solved the problem of service
providers being burdened with and constrained by too many separate contracts. But the city still had to ensure that service providers were working

toward the common goal of moving homeless people and families into permanent housing. HSD planned to do so through incentives in the contracts that reward service providers for meeting performance benchmarks.

Despite its intentions, however, the city can achieve only so much acting independently: not only does Seattle not provide social services directly, but it is just one of several social service funders in the region. King County (which encompasses Seattle) and the local United Way are also major funders of the same service providers. Even if the city designed contracts with clear performance measures, almost half of the service providers' work would still be commissioned by those two other funders; if they continued to promote separate objectives, social services would remain disjointed and ineffective. Thus, to effectively create a coherent agenda for every local service provider, Seattle had to align its goals with those of the other key stakeholders.

Over the course of a year—"I don't even want to think about how many meetings," Johnson remarks—the city, county, United Way, and political leaders formulated a common set of goals for homeless services that emphasize long-term desired outcomes. Chief among these is getting the homeless into permanent housing and preventing them from returning to the streets. Another key goal, because the African American and LGBTQ populations have historically been underserved, is ensuring that every homeless demographic receives services proportional to its needs. Finally, these stakeholders wanted service providers to collect more accurate and comprehensive data about their operations. All three social services funders now include performance incentives tied to these outcomes in their contracts.

With the new contracts in place, HSD has instituted monthly meetings with service providers to ensure that they achieve adequate progress toward these goals. In the past, providers were rarely monitored beyond ensuring that they complied with local regulations. Whether performing well or not, HSD had little insight into or influence over their activities. Now, says Johnson, "If providers are underperforming, every month there's going to be a discussion about how to remedy that." These conversations (aided by the newly collected data) have already helped the city and service providers align their resources and priorities—for example, by identifying families who are falling through the cracks and devising plans to provide them with tailored aid.

By creating flexible portfolio contracts, setting clear goals, and gathering better data, Seattle has vastly expanded its ability to decrease the local homeless population and mitigate the harms that homeless individuals face. "Now that they have the actual performance data, they can figure out what is working for people and what is not," Grover-Roybal explains, adding that HSD has already learned a great deal about how people are actually moving through services and which providers are most effective.

Although much work remains—more resources and new policies are necessary to address the underlying issues—these gains are beginning to translate directly to improvements in the lives of homeless people and families in Seattle: in the first quarter of 2018, more than 3,000 households were moved into permanent housing or maintained their housing through city investments in homeless services—a 69 percent increase over the first quarter of 2017.[30] In fact, only six months after HSD began piloting portfolio contracts and performance-based pay, Shakira Boldin's family was placed into permanent housing. "I can't really explain the feeling," she says. "Every day I wake up and I just feel blessed that me and my child have a roof over our heads. I feel like my future is bright."[31]

* * *

One of the smart city's greatest and most pernicious tricks is that it misappropriates the role and meaning of innovation. First, it puts innovation on a pedestal by devaluing traditional practices as emblematic of the undesirable dumb city. Second, it redefines innovation to simply mean making something more technological.

This chapter, which provides our most extensive look at what actually enables and sustains Smart Enough Cities, defies that logic: the most important innovations occur on the ground rather than in the cloud. Technological innovation in cities is primarily a matter not of adopting new technology but of deploying technology in conjunction with nontechnical change and expertise (of course, innovation need not involve technology at all). Cities must overcome numerous institutional barriers just to make data meaningful and actionable. MODA and DataSF did not need to find the optimal machine learning algorithm—instead, they had to painstakingly break down departmental silos, create new practices to manage data repositories, and train staff in new skills.

Seattle most clearly illustrates the benefits of recognizing that innovation means more than just "use new technology." Municipal governments operate within a remarkably complex structure: their powers and capabilities are limited, and they must engage with numerous other institutions.

Yet no smart city technologies are designed with any such structure in mind; a focus solely on technology would have left HSD powerless to improve homeless services. When Jason Johnson identifies contract reform as the city's "only tool" to address endemic homelessness and highlights the year of meetings required to unite several social services funders behind shared goals, we see clearly that technology is impotent to address many of the pressing challenges that cities actually face. Data is helping Seattle to evaluate programs and identify where resources are needed, but it would have little impact without these more systemic reforms.

Technology also cannot provide answers—or even questions—on its own: cities must first determine what to prioritize (a clearly political task) and then deploy data and algorithms to assess and improve their performance. *Wired* famously promised that "the data deluge makes the scientific method obsolete" and represents "the end of theory,"[32] but in today's age of seemingly endless data, theory matters more than ever. In the past, when they collected minimal data and had little capacity for analytics, cities had few choices about how to utilize data. Now, however, cities collect extensive data and can deploy cutting-edge analytics to make sense of it all. The magnitude of possible analyses and applications facing them is overwhelming. Without a thorough grounding in urban policy and program evaluation, cities will be bogged down by asking the wrong questions and chasing the wrong answers. Seattle had lots of data about homeless services, for example, but lacked a strategy to guide data collection and analysis toward its ultimate goals.

"The key barrier to data science is good questions," observes Joy Bonaguro. Using data effectively in city governments requires determining which issues, among the many that cities confront, can effectively be addressed with data. Furthermore, improving operations with data often hinges not on developing a fancy algorithm but on thoughtfully implementing an algorithm to serve the precise needs of municipal staff. Bonaguro therefore seeks far more than technical expertise when building her team. "When we hire data scientists," she explains, "I really want someone who does not want to just be a machine learning jockey. We need someone who is comfortable and happy to use a range of techniques. A lot of our problems aren't machine learning problems."

In Chicago, Chief Data Officer Tom Schenk has the same priorities when hiring for his team. "The challenge is trying to find the data scientist and the researcher who can work well with departments, because that's the key aspect," he notes. "And there are a lot of researchers who are not great about that. We need to find one that can come in and not just do the statistics, but also sit in a room with a manager and find out all the information they need to know."[33]

One of Chicago's data science projects relied on precisely this kind of on-the-ground research and relationship building. Several years ago, Schenk began working with the Chicago Department of Public Health (CDPH) to proactively identify the local food establishments that posed the greatest threats to public safety. If Schenk could predict which restaurants were most likely to violate public health regulations—precisely the type of task that machine learning excels at—he could direct food safety inspectors (known as "sanitarians") to those locations and help CDPH make the best use of its limited resources.

On a technical level, the project sounded easy: develop a machine learning model that references historical food inspections to identify indicators of unsafe establishments. But Schenk knew the project would not be so simple: he needed to study a large city agency with complex operations about which he knew little, and then develop and deploy an algorithm that could be embedded in its daily operations. So instead of focusing solely on how to create the most sophisticated algorithm, Schenk prepared himself for intense research.

When Schenk approached CDPH's food inspections manager about the project, he emphasized that a successful collaboration required him to gain a deep understanding of CDPH's goals and operations. "We're going to ask you a lot of questions that are going to seem very rudimentary," Schenk told her at the time. Such research is vital to successfully using data science in government, he explains. "It's super easy to miss what departments don't think is important because it's very banal to their process, but is key for our statistical modeling. The stats aren't hard for us. We spend most of our time talking to the client, trying to understand everything so we can apply statistics."

Even once the machine learning model appeared to be operational, however, it still had to undergo another critical step: experimental evaluation. With a background in policy analysis and medical research, Schenk knew it was important to test every model before deploying it. "We know there can be a disconnect between the logic we think is correct, and what happens in the real world," he says. "We need to introduce experiments to make sure things actually work."

Schenk designed a double-blind experiment to evaluate whether his algorithm could actually help sanitarians catch a greater number of "critical violations" such as a failure to heat or refrigerate food at the proper temperature. On the basis of internal tests and simulations, he expected that his machine learning approach would drastically increase the department's efficiency at finding critical violations. But the experiment indicated that the algorithm produced only a negligible improvement. "It took us a long time for us to dig into it and find out what was going on," Schenk recalls.

With such a large gap between expectations and reality, Schenk realized that despite his best efforts, he must have overlooked a key aspect of the food inspections process when designing the algorithm. He went back to CDPH's food inspections manager to determine what he had missed. During their conversation, she mentioned in passing that all of the sanitarians had just been reassigned to new neighborhoods for the first time in several years. This was the clue Schenk was seeking: he realized that what had appeared to be an important factor in predicting violations— zip codes—was actually just a reflection of differences between sanitarians, as each assigned food violations according to slightly varying standards. Schenk had not been aware that sanitarians were assigned to specific zip codes, and so did not account for this in his modeling.

Although disappointed by the algorithm's failure during this trial, Schenk saw the experiment as a success: it had uncovered a gap between the assumptions embedded in the model and CDPH's actual practices. Schenk updated the algorithm and several months later ran another experiment. This time, the improvement was notable. A simulation found that using the predictive model, CDPH could have improved its early detection of critical food safety violations by 26 percent. If CDPH had been following the recommendations of the predictive model, it would have discovered each critical violation a week earlier on average.[34] With these encouraging results in hand, Schenk and CDPH were finally confident that the model was ready for deployment. It has been in action, guiding where sanitarians conduct inspections, since 2014.

The Mayor's Office of New Urban Mechanics (MONUM) in Boston is developing an even sharper focus on science and research. "For many years now, we've been talking about the need to become data-driven, and that is clearly one important direction that we need to explore further," says MONUM co-founder and co-chair Nigel Jacob. "But there's a step beyond that. We need to make the transition to being science-driven in how we think about the policies that we're deploying and the way that we're developing strategic visions. It's not enough to be data mining to look for patterns—we need to understand root causes of issues and develop policies to address these issues."[35]

In April 2018, MONUM released a "Civic Research Agenda" comprising 254 questions, the answers to which will inform the city's efforts to improve life for all Bostonians. These questions range from big ("How can we gain a holistic understanding of the kind of future people want for Boston?") to small ("What can be done to lower the cost of construction?"), from technological ("How does technology play a role in perpetuating or addressing longstanding inequities across our city?") to nontechnological ("What is at the root of community opposition to new housing?").[36]

All of this is necessary, says Kim Lucas, MONUM's civic research director, to ensure that municipal projects are based on evidence and demonstrated civic needs. "You can't solve a real-life issue if you don't understand it, if you don't ask the right questions, and if you don't understand how to get the right information," she explains. "That's all research is: asking a question and then finding out the right information. And the next step when you have a finding is to do something with it."[37]

Staying grounded in research helps Boston avoid the perils of tech goggles. "Technology is a great tool, but it is not the answer," says Lucas. "Technology is a tool toward getting the answer more efficiently." Lucas relies on research, in other words, to "find the right tool to answer the right questions. If you're not asking the right question in the first place, how do you know that technology is the right approach? It may or may not be."

Which brings us back to the core message of this book: cities are not technology problems, and technology cannot solve many of today's most pressing urban challenges. Cities don't need fancy new technology—they need to ask the right questions, understand the issues that residents face, and think creatively about how to address those problems. Sometimes technology can aid these efforts, but technology cannot provide solutions on its own.

While that observation may appear obvious by now, confidence in the possibility of understanding and optimizing society through technical measures has been remarkably persistent not just over the past several years but over the past several centuries. The next chapter—the book's conclusion—will discuss the evolution of those beliefs. In exploring the similarities between past and present, along with how historical attempts to rationalize society have gone awry, it will demonstrate why smart cities are bound to fail. The book will conclude by highlighting how we can avoid such misguided and narrow thinking, synthesizing the lessons we have learned to provide a clear framework that can guide the development of Smart Enough Cities.

7 The Smart Enough City: Lessons from the Past and a Framework for the Future

We have, at this point, discredited many promises of the smart city and demonstrated the perverse effects of tech goggles. But perhaps some lingering doubt remains. After all, today's technology is remarkable in many ways: we can gather data about phenomena that were previously opaque, predict outcomes that once appeared unforeseeable, and interact with others on a historically unimaginable scale. Is it not possible, as Sidewalk Labs founder and CEO Dan Doctoroff asserts, that "digital technologies [will] bring about a revolution in urban life" on par with the upheavals introduced by the steam engine, the electric grid, and the automobile?[1]

Indeed it is. But just because a revolution stems from technology does not mean that its primary impacts will be technical: digital technologies designed for the smart city will drastically alter municipal governance and urban life. We must therefore be thoughtful about how cities adopt digital technology not for technological reasons—to ensure that they get the most advanced tools to maximize efficiency, for example—but because the technical infrastructure undergirding the smart city will go a long way toward determining the social and political infrastructure of twenty-first-century urbanism.

The most recent revolutionary technology that Doctoroff touts—the automobile—was remarkably destructive. Cars may not have been inevitably harmful to cities, but the attitude that cars were the key to urban progress was disastrous. Automobile manufacturers and oil companies promoted the "Motor Age," presenting corporate propaganda under the guise of providing socially optimal efficiencies. In the belief that smoother traffic was universally desirable, cities were reconstructed to support the efficient flow of vehicles at the expense of all else. They have spent much of the past half century attempting to undo those mistakes.

Today, technology companies promote the smart city as a new form of corporate propaganda. And if, as Doctoroff promises, digital technology produces a revolution similar to the one generated by cars—if the age of the smart city is anything like the Motor Age—it would be a travesty that mutilates cities.

The Motor Age is just one instance of the damage produced when utopians wearing tech goggles, fixated on the new science or technology *du jour*, reduce complex social and political matters to technical ones that can be optimized. In fact, despite the extent to which the smart city is presented as *sui generis*, today's discourse surrounding smart cities echoes the same beliefs and values espoused in the past, most notably in twentieth-century high-modern urban planning. Presaging the smart city, proponents of that movement believed that recent scientific and technological developments provided the tools to solve pressing urban issues. And in striving to optimize cities according to this narrow vision, high modernists perverted what they were trying to fix. This is not a historical anomaly but the inevitable result of reforms inspired by tech goggles, regardless of the specific technology: optimizing society with rationality and efficiency in mind requires reducing complex ecosystems to simplified schemas, often causing irrevocable damage.

The discredited tropes and schemes of high-modern urban planning rear their undying heads in the smart city, placing the future of urbanism at risk. The issue is not that today's data and algorithms are inherently flawed or malicious—just as earlier technologies and scientific methods were not inherently flawed or malicious—but rather that ecological systems such as cities are far too complex to perfectly rationalize and that attempts to do so often create long-term damage. We need not fear technology in general—but if history is any guide, we must be wary of those who promote bold visions of science and technology as providing solutions that transcend history and politics to produce an optimal society. History has told us that the world created under the influence of tech goggles is an undesirable one. We must instead pursue an alternative vision that bears no imprint of tech goggles: the Smart Enough City.

* * *

Germany at the turn of the nineteenth century provides an instructive example of the limits of hyper-rationalized planning. When a wood shortage threatened the economy, public officials began closely managing local forests to maximize wood production. New mathematical techniques helped scientists keep track of the environment and calculate, from each tree's size and age, the amount of wood that it would produce.[2]

The natural complexity of forests got in the way, however: the unsystematically scattered trees were hard to measure and the abundance of other wildlife drained resources that could otherwise help the trees grow larger, faster. Refusing to be thwarted in their quest to optimize timber yields and increase profits, the Germans undertook a massive effort to cultivate more rational and manageable forests. They cleared existing growth and planted new trees in long, ordered rows, simplifying the natural environment down to its most essential, timber-producing components. Where once lay underbrush and a disordered mix of all ages and types of trees soon grew manicured rows of uniform trees along open pathways.

The results were, at first, remarkable. Wood production skyrocketed, bolstering the economy, and German forestry practices spread across the world. The forests' visual order became synonymous with their underlying bureaucratic order. But after one or two generations of trees had grown in any given forest—roughly a century—production severely declined. Some forests died completely.

As recounted by the political scientist James C. Scott in *Seeing Like a State*, this tale reveals that the downfall of Germany's forests was caused not by some unexplainable ecosystem collapse but by an early version of the tech goggles cycle. First, tech goggles: inspired by new scientific methods of mathematical analysis, German officials were confident that making forests more measurable and manipulable was the key to enhancing timber production. Next, technology: the Germans implemented this vision by transforming forests from unmanageable and incomprehensible thickets into regimented factories for commercial timber. And finally, reinforcement: as these practices gained worldwide esteem, the perspective of tech goggles became entrenched through new social conceptions that commodified "nature" as "natural resources." For example, writes Scott, "the actual tree with its vast number of possible uses was replaced by an abstract tree representing a volume of lumber or firewood."[3]

Germany's myopic approach to optimizing tree growth ignored and devalued elements that proved essential—quite literally, it missed the forest for the trees. Creating rational forests optimized for tree growth entailed not merely excluding bushes, plants, birds, and insects from scientific models of timber production but eliminating them from forests almost entirely. Such elimination is the inevitable result of reconstructing the world according to narrow visions: what does not get measured is dismissed as unnecessary and detrimental. And while at first these practices appeared to spur more efficient tree growth, they forever reduced the forests to brittle monoculture environments that lacked the biodiversity to maintain nutrient-rich soil and protect against devastation from disease and bad weather. Despite extensive efforts, the Germans were unable to fully revive these forests.

For Scott, the parable of the forests "illustrates the dangers of dismembering an exceptionally complex and poorly understood set of relations and processes in order to isolate a single element of instrumental value."[4] From the narrow perspective of tech goggles, it may appear possible to optimize the element of interest. But it seems that way only because tech goggles simplify and distort a given ecosystem into something that can be optimized. Acting on this vision surely optimizes *something*, but it may not be what was intended and may generate unforeseen and irreparable damage. This is the fundamental danger of tech goggles, and one that we have seen rear its head several times.

Unfortunately, such myopic and unsuccessful schemes have not been limited to forests. The same destructive, reductionist ideas that drove the German foresters have since motivated many reformers striving to improve society.

A notable example of such thinking is the ideology of high modernism, which emerged following the incredible advances in science and technology in the nineteenth and early twentieth centuries. During those years, humans took flight, discovered relativity and quantum mechanics, brought electric power into homes, and invented the telephone and the internal combustion engine. Making possible an astonishing range of advances, from vaccinating populations to providing transportation over long distances, science and technology provided solutions to countless previously intractable problems.

High modernism was not driven by a general appreciation for science, however. Instead, as Scott describes, high modernists embraced a broad and grandiose faith in scientific reasoning, believing that "rational engineering of all aspects of social life [would] improve the human condition."[5] Most dangerously, they presumed that scientific knowledge granted them an authority to reform society that overrode all other forms of judgment. Now that scientific approaches could devise optimal solutions to social problems, high modernists believed, politics was no longer necessary: public deliberation and political interests would merely impede or distort the ideal solutions they could develop. In fact, many high modernists asserted that to fully realize their utopian visions, existing habitats would have to be abandoned—a new and optimal society could be achieved only by starting from a blank slate.

An essential attribute of high-modern acolytes, notes Scott, is that they "see rational order in remarkably visual aesthetic terms. For them, an efficient, rationally organized city, village, or farm was a city that *looked* regimented and orderly in a geometrical sense."[6] This disposition values not just any visual order but a particular order in which the world is observed from above. Scott attributes the growing appeal of this perspective in the twentieth century to the development of the helicopter and airplane. Looking down on the world from above as a god would, many high modernists felt omniscient and omnipotent.

Urban planning is one of the domains where high modernists were most influential, and where the limits of that ideology became most apparent. An early proponent of such thinking was the British urban planner Ebenezer Howard. In 1902, Howard published *Garden Cities of To-Morrow*, deploring the rise of "crowded, ill-ventilated, unplanned, unwieldy, unhealthy cities."[7] He proposed the Garden City as a new type of rational community where every facility was placed in its proper location. At the Garden City's center is a large garden, surrounded by core public buildings such as the town hall and library. Further out from the center, along Grand Avenue, are schools, playgrounds, and houses of worship. The outer ring of the town has designated locations for factories, warehouses, and other facilities.

Howard used mathematical formulas to precisely specify how to maximize social welfare in Garden Cities: the proper balance between housing and jobs; the need for amenities such as playgrounds, schools, and open space; and the optimal population. These formulas could even inform planners when the population of a particular Garden City surpassed its capacity (approximately 30,000–50,000 people) and a new, peripheral one needed to be developed several miles away.[8]

Howard believed that the Garden City necessitated a radical departure from the past. He saw little worth salvaging in London; instead, he maintained, "better results [can] be obtained by starting on a bold plan on comparatively virgin soil." Continuing to inhabit existing and obsolete cities when "modern scientific methods" promised a better way, Howard asserted, would be like clinging to disproven geocentric philosophies in stubborn resistance to modern astronomy.[9]

Ebenezer Howard was just the start, however. No one better illustrates the perspective and danger of high-modern urban planning than the Swiss-French architect Le Corbusier (born Charles-Édouard Jeanneret), who took Howard's utopian dream to even more extreme heights. Despising Paris as a "vision of Dante's *Inferno*," Le Corbusier in 1933 proposed the Radiant City: a "vertical garden city" that was to be "an organized [and] ordered entity."[10]

Among the many scientific and technological advances of his day, Le Corbusier was particularly inspired by the airplane. In his 1935 paean, *Aircraft*, he described the aerial perspective as "a new function added to our senses." And when Le Corbusier considered from above "the cities where it is our lot to be," he was unnerved: "The airplane indicts the city [as] old, decayed, frightening, diseased."[11] Le Corbusier thus saw no choice but to pursue a fresh start. "We must refuse to afford even the slightest consideration to . . . the mess we are in now. There is no solution to be found there," he wrote. "The only thing to do is to take a sheet of clean paper and to begin work on the calculations, the figures, the realities of life as it is today."[12]

Le Corbusier hailed his Radiant City as a place of "total efficiency and rationalization." He designed it, following a linear and Cartesian logic, to possess a rigid functional segregation that would replace the "artificial mingling of functions only indifferently related to one another" that was endemic in existing cities. In a style known as "towers in the park," neighborhoods would consist of skyscrapers surrounded by vast open spaces. Residences, factories, shopping centers, and other facilities would each be placed in designated sectors. Moreover, to reduce the inefficiencies that attend shopping and meal preparation, Le Corbusier envisioned centralized catering services that would deliver hot meals straight to people's doors.[13] He even proposed having factory workers live separately from their families to minimize transportation between the residential and industrial zones.[14]

All of this was, of course, to be optimized through modern scientific methods. Le Corbusier devised "prodigiously true" plans to determine the exact needs of residents—from living space to playgrounds to sunlight— and to allocate resources according to those needs. He declared that this approach had enabled him to produce the "correct, realistic, exact plan" for cities that "has taken account of nothing but human truths."[15]

To Le Corbusier, this meant that his plan was "incontrovertible" and thus transcended politics: "This Plan has been drawn up well away from the frenzy in the mayor's office or the town hall, from the cries of the electorate or the laments of society's victims. It has been drawn up by serene and lucid minds." Le Corbusier believed that the Radiant City represented the unique solution for an ideal society, and that no politician, law, or member of the public should be allowed to stand in the way of its creation.[16]

Although Le Corbusier was not provided with much opportunity to build cities himself, the places that were developed following his vision demonstrate the limits and dangers of high-modernist urban schemes.

Le Corbusier's dream of a utopian city founded on a blank slate was realized in the Brazilian capital of Brasília. Founded in 1960 on previously empty land, Brasília was designed in the style of the Radiant City by the

Figure 7.1
Brasília's South Wing, containing a grid of residential superquadra.
Source: Photograph by Eric Royer Stoner, "Escala residencial," Brasília, Brazil (August 2007).

architects Lúcio Costa and Oscar Niemeyer. The city followed strict spatial segregation that created distinct sectors for housing, work, recreation, and public administration. Self-contained *superquadra* (super blocks) contained apartment buildings and facilities such as schools and retail establishments, and were built in proportion to what were believed to be the "ideal" conditions for the population, such as the UNESCO standard of 25 square meters of green space per resident.[17]

But instead of being healthy and egalitarian, as Le Corbusier's calculations predicted of a Radiant City, Brasília was dull and dispiriting. As the anthropologist James Holston documents in *The Modernist City*, the conditions of the city "contradicted what was intended." Residents "coined the expression *brasilite*, meaning 'Brasíl(ia)-itis,'" to describe the trauma of living there. Unlike the previous capital, Rio de Janeiro, where streets and public squares served as "the points of sociality" replete with festivals, children playing, and adults mingling, Brasília was "a city without crowds."[18] And rather than produce equality, the city's design engendered merely anonymity. In fact, while elites appreciated Brasília's economic opportunities and high living standards, the laborers who built the city were shunned and subjugated by the government. Through political conflict and worker rebellion, Brasília became a city of extreme social and spatial segregation in which the majority of the population lived in unplanned, illegal settlements on the city's periphery.

The predilections of high-modern urban planning spread to the United States, finding their most notable champion in Robert Moses. Like Howard and Le Corbusier, Moses was motivated in large part by a desire for singular functionality and visual order. According to the biographer Robert Caro, Moses's "vast creative energies were fired by the vision of cleanliness, order, openness, sweep—such as the clean, open sweep of a highway—and were repelled by dirt and noise, such as the dirt and noise he associated with trains."[19] In part for this reason, Moses built countless parkways while strongly resisting all efforts to enhance public transit. And believing, as did Le Corbusier, that his plans were socially optimal and thus transcended traditional forms of public decision making, Moses was notorious for ignoring public input and using dirty, heavy-handed tactics to implement his visions.

Moses also oversaw New York City's urban renewal efforts. Much of the public housing constructed under Moses (as in many other U.S. cities) was built in Le Corbusier's "towers in the park" style. Although underfunding and political neglect were also to blame, the Brasília-esque design of these complexes (such as the Fort Greene Houses in Brooklyn) contributed to their becoming what the journalist Harrison Salisbury called "the new ghettos" and "human cesspools."[20] Moreover, these supposedly benevolent projects often provided cover for the mass relocation of lower-income and black residents,[21] leading James Baldwin to declare that "urban renewal . . . means negro removal."[22]

The influence of tech goggles permeates the beliefs and designs of Howard, Le Corbusier, and Moses. All three possessed an excessive faith in order and efficiency that led them to distort the nature of urbanism and reject democracy. They saw themselves as solving technical problems with objective answers rather than making political decisions that involve complex trade-offs and that could engender legitimate differences of opinion. Le Corbusier in particular believed that he had developed the one and only solution for human existence, never recognizing that the question of what should be efficient (let alone whether efficiency is even a worthwhile goal) is a normative one.

This ideology explains why the cities and developments fashioned according to high-modern dreams failed to create livable and equitable urban environments: planners relied on the misguided perception that the value of cities comes from their rational organization and their ability to efficiently provide goods and services, and unwaveringly designed cities to maximize these ends. But just as optimizing the German forests for timber production required eliminating most plants and animals—much of what makes a forest a forest—so creating ideal cities according to high-modern visions required stripping away mixed uses, crowds, and tradition—much of what makes a city a city. That Brasília and New York were plagued by political conflict in addition to design flaws merely emphasizes what high modernists dangerously overlook: clever technical plans cannot eliminate politics.

Diagnosing these schemes as "the sacking of cities," Jane Jacobs in 1961 criticized top-down, superficially rational planning in *The Death and Life of Great American Cities*. Animating Jacobs's excoriation was high modernism's misunderstanding of "[t]he kind of problem a city is." By rejecting the preeminence of visual, planned order and instead valuing the lived experiences of urban residents, Jacobs recognized that what high-modern planners perceived as "chaos" and "disorder in the life of city streets" actually "represent a complex and highly developed form of order." She identified urban dwellers not as abstract agents who merely require efficient food delivery and a scientifically determined quantity of parks and housing, but as people engaged in "intricately interconnected, and surely understandable, relationships."[23]

Jacobs concluded that cities are not problems of "simplicity" or "disorganized complexity" that could be solved using the types of equations pioneered over the previous two centuries. Instead, she saw cities as ecosystems of "organized complexity" replete with countless interrelated components—environments that modern mathematical methods were ill-equipped to systematize or optimize.[24]

Yet because the tech goggles of the day were forged in the fires of those methods, high modernists like Le Corbusier and Moses could not perceive what fell beyond the purview of their mathematical schemas. Their deeply simplified notions of urbanism led them to stamp out the features that actually foster vibrant and buoyant communities and to create what Jacobs called the "anti-city."[25]

* * *

When we keep the lessons of history in mind, the smart city no longer appears to represent a bright new future—instead, it signals a regressive return to an ideology that has already been pursued and condemned. Even the language highlighting today's visions is remarkably evocative of the past: just as Le Corbusier hailed the Radiant City as "harmonious and lyrical," now MIT's Senseable City Lab describes its intelligent intersections that will obviate streetlights as "orchestra conductors" that will help "lanes of cars merge harmoniously."[26] One cannot help concluding that if Le Corbusier were alive today, he would be one of the smart city's most vocal and provocative boosters.

The primary distinction between high-modern urbanism and the smart city is that our tech goggles have evolved, inspired by the advancements in science and technology over the past several decades. Whereas utopian visions of twentieth-century cities prioritized visual order, drawing on new capabilities of flight to conceive of the city from above, utopian visions of smart cities prioritize digital order, drawing on new capabilities of data collection and analysis to conceive of the city from a computer. Last century's acolytes believed that new methods from the physical sciences would solve all social ills; today's place their faith in the smart city's holy trinity: Big Data, machine learning, and the Internet of Things.

To put the problem in the terms of Jane Jacobs, those looking through tech goggles are once again misdiagnosing "the kind of problem a city is." Rather than grappling with cities as problems of organized complexity or with the fundamental social and political challenges of urban life, smart city idealists describe cities as abstract technical processes that can be optimized using sensors, data, and algorithms. Blind to the countless aspects of urbanism that cannot be reduced to an app or an algorithm, they risk creating a modern incarnation of the anti-city.

The technology company Hitachi, for example, describes cities as "convenient places to live since they are equipped with social infrastructure such as electricity, water, and public transport, along with a variety of facilities such as housing, offices, and commercial facilities."[27] Living PlanIT, another company building smart city technologies, insists that "[c]ities need an 'operating system.'"[28] The startup accelerator Y Combinator, a

Figure 7.2
A diagram created by Sidewalk Labs depicting a computational vision of the city. Blue lines emanate from sensors on buildings and cars to represent their perceptions, while abstract boxes (containing the only solid lines in the figure) mark people and cars. The digital order presented in this utopian vision of the smart city echoes the visual order of twentieth-century utopian visions (figure 7.1).
Source: Sidewalk Labs, "Vision Sections of RFP Submission" (October 17, 2017), p. 71, https://sidewalktoronto.ca/wp-content/uploads/2017/10/Sidewalk-Labs-Vision-Sections-of-RFP-Submission.pdf.

company at the vanguard of Silicon Valley's tech industry, puts it most succinctly, writing that the primary question in its effort to build smart cities is, "What should a city optimize for?"[29]

Another statement from Y Combinator is remarkable for the precision with which it describes how cities appear through tech goggles: "Our goal is to design the best possible city given the constraints of existing laws."[30] The sentence is constructed like a typical mathematical optimization problem, demonstrating how the company approaches smart cities from an engineer's perspective. The expressed desire to create "the best possible city" reveals Y Combinator's myopic belief that an objectively optimal city exists, overlooking the politics, history, and culture of cities, as well as the diverse and often conflicting wants and needs of urban residents (indeed, it is hard to imagine that Y Combinator, a company steeped in Silicon Valley riches, shares a definition of "best" with the many communities being displaced by that wealth). Finally, Y Combinator's identification of "existing laws" as the sole constraint hampering cities highlights the company's incredibly narrow and shortsighted understanding of urban challenges and progress. By its logic, any flaws in existing cities have nothing to do with resource limitations or social conflict—just laws based on obsolete models of managing information and resources. The company's scorn for laws echoes that of the previous generation of utopian city building, in which Le Corbusier proudly declared that his plan "has ignored all current regulations."[31]

This brings us to the strongest indicator that technophiles perceive cities as little more than abstract staging grounds for efficient mobility solutions and service delivery: the persistent desire of technologists to build smart cities from scratch. Despite the significant challenges faced by early smart city efforts such as Masdar City in the United Arab Emirates and Songdo, South Korea, both of which remain largely desolate,[32] not to mention the similar failures of earlier tabula rasa cities such as Brasília, today's technologists have heedlessly taken up the cause. Y Combinator eagerly proclaims that "it's possible to do amazing things given a blank slate."[33]

Sidewalk Labs takes this perspective further, asserting that it is not just possible but in society's best interest to build smart cities without the encumbrances of history or place. Dan Doctoroff has stated that "there is an inverse relationship between your capacity to innovate, and the actual existence of people and buildings."[34] The company boldly acted on this vision in 2017, announcing a partnership in Toronto to develop an undeveloped

12-acre waterfront parcel into "the world's first neighbourhood built from the internet up."[35]

The most pernicious aspect of such ambitions is that they are presented as value-neutral and universally desirable. In conceiving of cities as optimization and technology problems and observing only what can be made efficient, technologists come to equate technical solutions with socially optimal ones. Such thinking diminishes their appreciation for the multiplicity of perspectives and needs that exist in (and perhaps cultivate) cities. Just as Le Corbusier believed that his "incontrovertible" plan had "taken account of nothing but human truths,"[36] Sidewalk Labs promises that their new city, with "ubiquitous connectivity designed into its very foundation," would be "a place that gives people more of what we love about cities with less of what we don't"[37]—somehow presuming in defiance of all history that there exists an obvious, single model of what a city should be and that technology can immunize cities from the intractable challenges of urban governance and life.

Yet even in its first year of development, Sidewalk Labs' Toronto project has been beset by political disputes. Recognizing that digital kiosks (like those of LinkNYC) would be just the tip of the data collection iceberg in a city neighborhood run by Sidewalk Labs, many Toronto residents have demanded more information about what data will be collected and how it will be used. The project is also haunted by the specter of shifting the management and ownership of public services over to an unelected and unaccountable private company, whose efforts will be aided by cuts in regulations. Sidewalk Labs has preached their intentions to include the community in developing this project. But after several public meetings at which the company released scant details about its plans, one local technologist declared that "the public engagement process is off the rails."[38] Moreover, the project's rapid timeline makes it almost impossible for the public to exert a meaningful influence over its development. Sidewalk Labs' Toronto neighborhood may indeed represent a new type of urbanism with technology at its core, but not one with "more of what we love about cities." Instead, it will be rife with the same issues that have plagued countless other cities: unaccountable decision making, privatized public services, and eviscerated political debate.

Simply put, utopian technological solutions fail to provide the answers that cities need.

"There's a real danger with some of this stuff that I call 'smartwashing' a problem that actually needs real investment," remarked Boston's chief information officer Jascha Franklin-Hodge at a 2017 conference about the Internet of Things. "We say, 'Let's just throw some technology fairy dust at it and that's going to make it go away.' But often the question of how that technology is going to deliver real and meaningful outcomes for constituents is not really answered."[39]

Nigel Jacob, Franklin-Hodge's colleague in Boston, shares a similar frustration. He recounts, "We've had so many conversations with vendors where they show up with the big pitch: 'We'll solve every problem in your city, if you buy this one technology.'" Every time, Jacob says, Boston would explain why it was not bullish on the proposed technology. But the objections were rarely heeded. "Sometimes the company would come back with a better pitch, but more often they would not. They would go to another city that asked fewer questions."[40]

Exasperated, Jacob and his team put their most common feedback into a single document that they could share with companies and technologists. In September 2016, they released the "Boston Smart City Playbook," espousing Boston's intention to deploy technology that is "people-centered, problem-driven, and responsible."[41]

"So far," the playbook begins, "many 'Smart City' pilot projects that we've undertaken here in Boston have ended with a glossy presentation, and a collective shrug. Nobody's really known what to do next, or how the technology and data might lead to new or improved services." The Playbook then lists six frank recommendations, starting with

1. "Stop sending sales people: . . . send us someone who knows about cities, someone who wants to walk in the shoes of our City workers or talk to residents";
2. "Solve real problems for real people: . . . we can't help feeling like this keeps getting lost. . . . How do you know that 'the problem' you're addressing really is a problem?"; and
3. "Don't worship efficiency: . . . focusing on 'efficiency' assumes that we've already figured out what services to deliver to residents, and now just have to make it all cheaper. That's unfortunately not the case."[42]

The playbook demonstrates Boston's sense that the smart city is more of a distraction than a goal. "We are trying to be very values focused, and in particular to solve real problems for real people," explains Franklin-Hodge. "Our smart cities strategy is just our city strategy. It's about having an equitable city. It's about having economic development. It's about sustainability. If I'm developing a smart city strategy that isn't directly tied to those real needs and challenges we have as a city, then I'm not doing my job."[43]

Boston's commentary highlights the need for technologists to ground themselves in the needs of city governments and urban residents. Just as Jane Jacobs understood that a city should be designed according to the lived experience of its inhabitants rather than urban planners' top-down, visual notions of order, Jacob and Franklin-Hodge recognize that cities should adopt technology to meet the genuine needs of urban residents rather than to follow engineers' computational notions of order.

Digital technology is not inevitably harmful. But the view through tech goggles, that it is possible to create optimal cities using new technology, diverts attention away from and subverts opportunities to democratically and equitably improve cities. In its very name, the smart city positions being "smart" as the goal—as if better data and technology inherently benefit society—leading to an urban agenda of enhancing technology without fully considering the implications of doing so or the variety of alternative goals that could be pursued. When Y Combinator asks, "What should a city optimize for?," it presupposes that optimization is the primary tool with which to improve urban life, ignoring the multitude of issues that are not reducible to an optimization problem. This perspective tends to entrench the status quo and to hinder other, more important reforms.

The Smart Enough City reorients this logic by prompting a fundamental question: smart enough for what? In the Smart Enough City, where being "smart" is a means rather than an end, the focus can rightfully turn to the social needs that technology addresses. As Boston demonstrates, that means employing technology only as it is able to alleviate "real problems for real people."

This is the essential paradigm shift for cities, both in the United States and beyond. Although this book has focused on developments within the United States, many of the same trends, opportunities, and challenges exist around the world. Singapore is eagerly deploying autonomous vehicles and welcomed the world's first self-driving taxis.[44] In Ethiopia, Addis Ababa has deployed a "smart parking system" to address the city's severe shortage of parking.[45] Participatory budgeting was born in Brazil, where to this day it is used in hundreds of cities (often comprising larger portions of municipal budgets than in the United States);[46] Brazil is a leader in synthesizing online and offline approaches to participatory budgeting.[47] In China, the city of Xinjiang has aggressively deployed a predictive policing platform that draws on a great deal of personal data,[48] leading the *Wall Street Journal* to deem it "one of the most closely surveilled places on earth."[49] In 2017, London began rolling out InLinkUK kiosks (an almost exact replica of the LinkNYC program) with minimal public outreach,[50] and it has placed countless sensors in metro stations to track the behavior of commuters.[51] Barcelona has spearheaded several Internet of Things deployments across the city,[52] all while developing participatory processes to curb the power of tech companies, to provide transparency regarding the use of algorithms, and to transfer ownership and control of data to the public.[53] Through these efforts and countless more, cities across the globe are stumbling and pioneering their way toward new uses of technology. We all bear the responsibility for pushing these efforts toward the ideals of smart enough, rather than smart, cities.

* * *

From prenatal healthcare in Columbus to participatory budgeting in Vallejo to proactive social services in Johnson County to the surveillance ordinance in Seattle to data drills in New York City, we have seen numerous ingredients for creating Smart Enough Cities. To support and further these efforts, I have summarized five essential principles for Smart Enough Cities that have emerged throughout the book. Although surely incomplete, this list will, I hope, help set forth an agenda for more livable, democratic, just, responsible, and innovative cities.

1. Address complex problems rather than solve artificially simple ones

Simplistic conceptions of social and political challenges always accompany tech goggles. The histories of German forests, high-modern urban planning, and the Motor Age demonstrate the destructiveness of this perspective: overlooking or striving to eradicate the world's natural complexity leads to "solutions" that address artificial problems and often create more problems than they solve.

Unfortunately, these same simplistic notions pervade the widespread hopes and dreams surrounding smart cities. Self-driving cars appear poised to create urban utopias, for example, but only because technologists focus disproportionately on efficient car travel as the hallmark of a good city. In their failure to recognize the many challenges and trade-offs related to transportation, or even the need to balance smooth traffic against other goals, technologists overestimate the benefits of automated vehicles while also ignoring other types of reforms. In other words, they oversimplify the problem of transportation into one that can be solved via optimization and then propose an elegant solution.

In contrast, Smart Enough Cities more fully grasp the complexity of urban issues and hence better recognize the limits and opportunities of technology. Rather than focusing on mobility as a matter merely of convenience, the Smart Columbus team recognized that mobility is interconnected with other challenges such as inequality. It further avoided the trap of artificial simplicity by engaging with a diverse local population to determine what transportation barriers they actually face. Doing so enabled Columbus to move beyond some of its simplistic initial notions and develop effective mobility reforms that address the real problems residents face. As Carla Bailo explains, "We really needed to look at it from a more holistic viewpoint."[54] Columbus cannot remove obstacles to mobility or equity with a single technology or policy reform, but its efforts will alleviate some the daily challenges that residents face.

2. Implement technology to address social needs and advance policy, rather than adapting goals and values to align with technology

This is at the heart of Franklin-Hodge's mantra for Boston: your smart city strategy should be the same as your broader city strategy. Smart Enough Cities are driven by clear policy goals and long-term planning efforts. They often embrace technology as a tool to advance their values, but the technology never dictates those objectives.

Tech goggles (and by extension the tech goggles cycle), on the other hand, shape urban innovation according to the logic and capabilities of technology. To address today's challenges in civic engagement and democracy, city governments and technologists have proposed countless technologies: online platforms, social networks, and 311 apps, all with the express purpose of making politics and governance simpler and more efficient. But power and politics are not optimization problems—being "smart" will not solve democracy. For example, 311 apps may make it easy to notify the government about a broken streetlight, but they do little to empower residents or generate deeper community ties.

Smart Enough Cities instead lead with social and political goals and deploy technology only to advance that agenda. They are not seduced by technologies that sound attractive but do not align with their plans and values. In contrast to typical civic engagement apps, online platforms such as Community PlanIt can help develop civic relationships and capacities by embracing "meaningful inefficiencies." Similarly, Vallejo altered local democratic practices by implementing participatory budgeting; new apps are engaging more people in the process, but the program's fundamental change came from giving the public new deliberative opportunities and decision-making power.

3. Prioritize innovative policy and program reforms above innovative technology

Smart Enough Cities create their most significant impacts through policy and process reforms that thoughtfully address local needs. Technology can make these reforms more effective, but it is never the driving force. In fact, many of the success stories we examined involve relatively simple data analyses and technologies, deployed to support innovative policies. Their deployment was successful because enhancing technology is just one form of innovation—a good program that relies on simple technology is better than a bad program that uses cutting-edge technology.

In smart cities, however, technology takes center stage. Eager to be seen as innovative and race-neutral, police departments have enthusiastically adopted predictive policing software. But they are missing the point: communities need a fundamental reshaping of police practices and priorities, not an enhancement by algorithm of the same old behavior. In fact, by providing a sheen of neutrality, predictive policing algorithms justify and exacerbate discriminatory inequities and police practices, thereby putting more systemic reforms further out of reach.

Smart Enough Cities should instead follow what I call the "limited tech test." When considering the use of a new technology, city leaders should ask the following questions: If it were possible to achieve the same outcomes without technology, would it still be innovative? Would the impacts be desirable? Smart Enough Cities adopt technology only when they can confidently answer in the affirmative. Striving to reduce incarceration and improve social services, for instance, Johnson County began providing aid to individuals suffering from mental illness to improve their lives and keep them out of the criminal justice system. Johnson County generated these benefits not by discovering a new, infallible algorithm to optimize and legitimize existing police practices, but by reforming its programs to address community needs and then increasing this program's efficacy with machine learning.

4. Ensure that technology's design and implementation promote democratic values

Tech goggles make it appear that complex social issues are technical problems for which technology can provide value-neutral and socially optimal solutions. This simplistic assessment leads to smart city technologies that are designed to enhance efficiency at any cost, with little assessment of their broader social impacts.

Many smart city technologies make governments and companies more efficient by collecting as much data as possible, a process that entails infringing on people's privacy and autonomy. Similarly, many smart cities operate with the aid of opaque and proprietary algorithms that are developed and deployed without public input. These trends create massive information and power asymmetries that empower governments and companies over those they track and analyze. In this way, the smart city is a covert tool for increasing surveillance, profits, and social control.

Embracing their role as public stewards to ensure that new technology benefits everyone, Smart Enough Cities reject the false dichotomy between smart and dumb cities that would have them eagerly deploy every new tool; they instead consider a broad range of designs for new technology to ensure that both the means and the ends support democracy and equity. Seattle and Chicago demonstrate that respecting and protecting individual privacy enables, rather than hinders, the deployment of new technology to improve urban life. Similarly, the algorithm task force in New York and surveillance oversight ordinance in Seattle demonstrate a clear path for municipalities to reverse the trend toward becoming black-box cities. As these examples indicate, rejecting or altering technology because it violates important values is not anti-technology—it is pro-democracy.

5. Develop capacities and processes for using data within municipal departments

It is easy to believe that technology can improve government simply by virtue of its sophistication. The reality is far more complicated: poor data quality limits analyses, siloed departments struggle to share data, and many departments have little trust in data to solve their problems. What makes data most useful is not having the most advanced technical capabilities but lowering institutional barriers and identifying the problems that data can address.

Municipal leaders such as Amen Ra Mashariki in New York City and Joy Bonaguro in San Francisco demonstrate how to deploy data to improve local governance—not by expecting data to magically optimize government or solve local issues but by building relationships with departments, fostering best practices for maintaining and sharing data, and training city staff in how to use data to improve their operations.

Smart Enough Cities should follow their lead, rejecting smart city rhetoric that prescribes newer and more advanced technology as the way for city governments to quickly solve every problem. They must instead focus on the painstaking work of developing the infrastructural (even quotidian) processes and practices that make data actionable.

* * *

New technology alters perceptions not just of what is possible but also of what the world can—and should—look like. Digital and data-driven technologies, accompanied by the widespread adoption of tech goggles, have convinced many that smart cities are what the challenges of the twenty-first century require: that smarter cities—more connected, more optimized, more efficient—will be better cities.

This seductive logic generates severe misperceptions and subverts opportunities to truly improve urban life. Instead of addressing the real issues that cities face, smart cities present novelty solutions to poorly specified problems. The realization of those solutions looms as a crisis in urbanism: the smart city will be a place where self-driving cars strangle downtowns and debilitate public transportation, where democracy is reduced to sending pictures of potholes with an app, where police use algorithms to justify and perpetuate racist practices, where governments and companies surveil public space to control behavior.

But despite how often we are told that the age of the smart city is imminent and inevitable, a better future is possible. We can create livable cities, where simple mobility technologies mitigate inequality and enhance public health. We can create democratic cities, where communication technologies aid new participatory processes that empower the public. We can create just cities, where machine learning algorithms help communities aid vulnerable residents. We can create responsible cities, where new technologies are designed to support privacy and democracy. We can create innovative cities, where data science is paired with nontechnological reforms to improve municipal operations and social services.

We can create these Smart Enough Cities, if only we possess the wisdom to seek them. Throw away those tech goggles, once and for all, and let's get started.

Notes

Chapter 1

1. Kevin Hartnett, "Bye-Bye Traffic Lights," *Boston Globe*, March 28, 2016, https://www.bostonglobe.com/ideas/2016/03/28/bye-bye-traffic-lights/8HSV9DZa4qPC1tH4zQ4pTO/story.html.

2. Senseable City Lab, "DriveWAVE by MIT SENSEable City Lab" (2015), http://senseable.mit.edu/wave/.

3. Remi Tachet et al., "Revisiting Street Intersections Using Slot-Based Systems," *PloS One* 11, no. 3 (2016), https://doi.org/10.1371/journal.pone.0149607.

4. "Massachusetts Ave & Columbus Ave," Walk Score (2018), https://www.walkscore.com/score/columbus-ave-and-massachusetts-ave-boston.

5. George Turner, quoted in PredPol, "Atlanta Police Chief George Turner Highlights PredPol Usage," PredPol: Blog (May 21, 2014), http://www.predpol.com/atlanta-police-chief-george-turner-highlights-predpol-usage/.

6. New York City Office of the Mayor, "Mayor de Blasio Announces Public Launch of LinkNYC Program, Largest and Fastest Free Municipal Wi-Fi Network in the World" (February 18, 2016), http://www1.nyc.gov/office-of-the-mayor/news/184-16/mayor-de-blasio-public-launch-linknyc-program-largest-fastest-free-municipal#/0.

7. White House Office of the Press Secretary, "FACT SHEET: Administration Announces New 'Smart Cities' Initiative to Help Communities Tackle Local Challenges and Improve City Services" (September 14, 2015), https://obamawhitehouse.archives.gov/the-press-office/2015/09/14/fact-sheet-administration-announces-new-smart-cities-initiative-help; National League of Cities, "Trends in Smart City Development" (2016), http://www.nlc.org/sites/default/files/2017-01/Trends in Smart City Development.pdf.

8. United States Conference of Mayors, "Cities of the 21st Century: 2016 Smart Cities Survey" (January 2017), p. 4, https://www.usmayors.org/wp-content/uploads/2017/02/2016SmartCitiesSurvey.pdf.

9. John Chambers and Wim Elfrink, "The Future of Cities," *Foreign Affairs*, October 31, 2014, https://www.foreignaffairs.com/articles/2014-10-31/future-cities.

10. John Dewey, *Logic: The Theory of Inquiry* (New York: H. Holt and Company, 1938), 108.

11. Bruno Latour, "Tarde's Idea of Quantification," in *The Social After Gabriel Tarde: Debates and Assessments*, ed. Mattei Candea (London: Routledge, 2010), 155.

12. Horst W. J. Rittel and Melvin M. Webber, "Dilemmas in a General Theory of Planning," *Policy Sciences* 4, no. 2 (1973): 155 (abstract).

13. Evgeny Morozov, *To Save Everything, Click Here: The Folly of Technological Solutionism* (New York: PublicAffairs, 2014).

14. Langdon Winner, *The Whale and the Reactor: A Search for Limits in an Age of High Technology* (Chicago: University of Chicago Press, 1986), 19, 29.

15. Adam Greenfield, *Against the Smart City* (New York: Do Projects, 2013), 32–33.

16. Gordon Falconer and Shane Mitchell, "Smart City Framework: A Systematic Process for Enabling Smart+Connected Communities" (2012), https://www.cisco.com/c/dam/en_us/about/ac79/docs/ps/motm/Smart-City-Framework.pdf.

17. Samuel J. Palmisano, "Smarter Cities: Crucibles of Global Progress" (address, Rio de Janeiro, November 9, 2011), https://www.ibm.com/smarterplanet/us/en/smarter_cities/article/rio_keynote.html.

18. Alana Semuels, "The Role of Highways in American Poverty," *The Atlantic*, March 2016, https://www.theatlantic.com/business/archive/2016/03/role-of-highways-in-american-poverty/474282/.

19. New York Times Editorial Board, "The Racism at the Heart of Flint's Crisis," *New York Times*, March 25, 2016, https://www.nytimes.com/2016/03/25/opinion/the-racism-at-the-heart-of-flints-crisis.html.

20. Winner, *The Whale and the Reactor*, 23.

21. Theodore M. Porter, *Trust in Numbers: The Pursuit of Objectivity in Science and Public Life* (Princeton, NJ: Princeton University Press, 1995), 8.

22. Marshall Berman, "Take It to the Streets: Conflict and Community in Public Space," *Dissent* 33, no. 4 (1986): 481.

Chapter 2

1. Stephen Buckley, interview by Ben Green, April 7, 2017. All quotations from Buckley in this chapter are from this interview.

2. Ryan Lanyon, interview by Ben Green, April 13, 2017. All quotations from Lanyon in this chapter are from this interview.

3. National Center for Statistics and Analysis, National Highway Traffic Safety Administration, "Critical Reasons for Crashes Investigated in the National Motor Vehicle Crash Causation Survey," Traffic Safety Facts: Crash Stats, Report No. DOT HS 812 115 (February 2015), 1; National Center for Statistics and Analysis, National Highway Traffic Safety Administration, "2015 Motor Vehicle Crashes: Overview," Traffic Safety Facts Research Note, Report No. DOT HS 812 318 (August 2016), 6.

4. Daniel J. Fagnant and Kara Kockelman, "Preparing a Nation for Autonomous Vehicles: Opportunities, Barriers and Policy Recommendations," *Transportation Research Part A: Policy and Practice* 77 (2015): 175.

5. Fagnant and Kockelman, "Preparing a Nation for Autonomous Vehicles," 173.

6. Jeffrey Owens, CTO, Delphi, speaking in TechCrunch, "Taking a Ride in Delphi's Latest Autonomous Drive" (2017), https://www.youtube.com/watch?v=wWdVfG lBqzE.

7. Kinder Baumgardner, "Beyond Google's Cute Car: Thinking Through the Impact of Self-Driving Vehicles on Architecture," *Cite: The Architecture + Design Review of Houston* (2015): 41.

8. Senseable City Lab, "DriveWAVE by MIT SENSEable City Lab" (2015), http://senseable.mit.edu/wave/.

9. Remi Tachet et al., "Revisiting Street Intersections Using Slot-Based Systems," *PloS One* 11, no. 3 (2016), https://doi.org/10.1371/journal.pone.0149607.

10. Baumgardner, "Beyond Google's Cute Car," 41.

11. Lisa Futing, quoted in Sam Lubell, "Here's How Self-Driving Cars Will Transform Your City," *Wired*, October 21, 2016, https://www.wired.com/2016/10/heres-self-driving-cars-will-transform-city/.

12. Henry Claypool, Amitai Bin-Nun, and Jeffrey Gerlach, "Self-Driving Cars: The Impact on People With Disabilities," *The Ruderman White Paper* (January 2017), pp. 16, 18, http://secureenergy.org/wp-content/uploads/2017/01/Self-Driving-Cars-The -Impact-on-People-with-Disabilities_FINAL.pdf.

13. Ravi Shanker et al., "Autonomous Cars: Self-Driving the New Auto Industry Paradigm," *Morgan Stanley Blue Paper* (November 6, 2013), p. 38, https://orfe.princeton .edu/~alaink/SmartDrivingCars/PDFs/Nov2013MORGAN-STANLEY-BLUE-PAPER -AUTONOMOUS-CARS%EF%BC%9A-SELF-DRIVING-THE-NEW-AUTO-INDUSTRY -PARADIGM.pdf

14. Peter D. Norton, *Fighting Traffic: The Dawn of the Motor Age in the American City* (Cambridge, MA: MIT Press, 2011), 248.

15. General Motors, "To New Horizons" (1939), posted as "Futurama at 1939 NY World's Fair," https://www.youtube.com/watch?v=sClZqfnWqmc.

16. Norton, *Fighting Traffic*, 1, 7.

17. Trevor J. Pinch and Wiebe E. Bijker, "The Social Construction of Facts and Artifacts: Or How the Sociology of Science and the Sociology of Technology Might Benefit Each Other," in *The Social Construction of Technological Systems: New Directions in the Sociology and History of Technology*, ed. Wiebe E. Bijker, Thomas P. Hughes, and Trevor Pinch (Cambridge, MA: MIT Press, 1987), 27.

18. Norton, *Fighting Traffic*, 130.

19. George Herrold, "City Planning and Zoning," *Canadian Engineer* 45 (1923): 129.

20. Norton, *Fighting Traffic*, 106.

21. George Herrold, "The Parking Problem in St. Paul," *Nation's Traffic* 1 (July 1927): 48; cited in Norton, *Fighting Traffic*, 124.

22. J. Rowland Bibbins, "Traffic-Transportation Planning and Metropolitan Development," *Annals of the American Academy of Political and Social Science* 116, no. 1 (1924): 212.

23. Norton, *Fighting Traffic*, 130, 134.

24. J. L. Jenkins, "Illegal Parking Hinders Work of Stop-Go Lights; Pedestrian Dangers Grow as Loop Speeds Up," *Chicago Tribune*, February 10, 1926.

25. Norton, *Fighting Traffic*, 138, 1.

26. Norton, *Fighting Traffic*.

27. Alan Altshuler, *The Urban Transportation System: Politics and Policy Innovations* (Cambridge, MA: MIT Press, 1981), 27–28.

28. Angie Schmitt, "How Engineering Standards for Cars Endanger People Crossing the Street," *Streetsblog USA*, March 3, 2017, http://usa.streetsblog.org/2017/03/03/how-engineering-standards-for-cars-endanger-people-crossing-the-street/.

29. Peter Furth, "Pedestrian-Friendly Traffic Signal Timing Policy Recommendations," *Boston City Council Committee on Parks, Recreation, and Transportation* (December 6, 2016), p. 1, http://www.northeastern.edu/peter.furth/wp-content/uploads/2016/12/Pedestrian-Friendly-Traffic-Signal-Policies-Boston.pdf.

30. Robert A. Caro, *The Power Broker: Robert Moses and the Fall of New York* (1975; repr., New York: Random House, 2015), 515.

31. Caro, *The Power Broker*, 515.

32. *New York Herald Tribune*, August 18, 1936; cited in Caro, *The Power Broker*, 516.

33. Caro, *The Power Broker*, 518.

34. Anthony Downs, "The Law of Peak-Hour Expressway Congestion," *Traffic Quarterly* 16, no. 3 (1962): 393.

35. Anthony Downs, "Traffic: Why It's Getting Worse, What Government Can Do," Brookings Institution Policy Brief #128 (January 2004), 4.

36. Anthony Downs, *Still Stuck in Traffic: Coping with Peak-Hour Traffic Congestion* (Washington, DC: Brookings Institution Press, 2005), 83.

37. Gilles Duranton and Matthew A. Turner, "The Fundamental Law of Road Congestion: Evidence from US Cities," *American Economic Review* 101, no. 6 (2011): 2618.

38. David Metz, *The Limits to Travel: How Far Will You Go?* (New York: Routledge, 2012).

39. Senseable City Lab, "DriveWAVE by MIT SENSEable City Lab."

40. "Massachusetts Ave & Columbus Ave," Walk Score (2018), https://www.walkscore.com/score/columbus-ave-and-massachusetts-ave-boston.

41. Ken Washington, "A Look into Ford's Self-Driving Future," Medium: Self-Driven (February 3, 2017), https://medium.com/self-driven/a-look-into-fords-self-driving-future-5aae38ee2059.

42. John Zimmer and Logan Green, "The End of Traffic: Increasing American Prosperity and Quality of Life," Medium: The Road Ahead (January 17, 2017), https://medium.com/@johnzimmer/the-end-of-traffic-6d255c03207d.

43. John Zimmer, "The Third Transportation Revolution," Medium: The Road Ahead (September 18, 2016), https://medium.com/@johnzimmer/the-third-transportation -revolution-27860f05fa91.

44. Emily Badger, "Pave Over the Subway? Cities Face Tough Bets on Driverless Cars," *New York Times*, July 20, 2018, https://www.nytimes.com/2018/07/20/upshot/ driverless-cars-vs-transit-spending-cities.html.

45. Cecilia Kang, "Where Self-Driving Cars Go to Learn," *New York Times*, November 11, 2017, https://www.nytimes.com/2017/11/11/technology/arizona-tech-industry -favorite-self-driving-hub.html.

46. Daisuke Wakabayashi, "Uber's Self-Driving Cars Were Struggling Before Arizona Crash," *New York Times*, March 23, 2018, https://www.nytimes.com/2018/03/23/ technology/uber-self-driving-cars-arizona.html.

47. Jeff Speck, interview by Ben Green, April 14, 2017.

48. Jeff Speck, "Autonomous Vehicles and the Good City" (lecture, United States Conference of Mayors, Washington, DC, January 19, 2017), https://www.youtube .com/watch?v=5AELH-sI9CM.

49. David Ticoll, "Driving Changes: Automated Vehicles in Toronto," discussion paper, Munk School of Global Affairs, University of Toronto (2015), https://munkschool .utoronto.ca/ipl/files/2016/03/Driving-Changes-Ticoll-2015.pdf.

50. Ben Spurr, "Toronto Plans to Test Driverless Vehicles for Trips to and from Transit Stations," *The Star*, July 3, 2018, https://www.thestar.com/news/gta/2018/07/03/ toronto-plans-to-test-driverless-vehicles-for-trips-to-and-from-transit-stations.html.

51. U.S. Department of Transportation, "Smart City Challenge: Lessons for Building Cities of the Future" (2017), p. 2, https://www.transportation.gov/sites/dot.gov/files/ docs/Smart City Challenge Lessons Learned.pdf.

52. Jordan Davis, interview by Ben Green, May 10, 2017. All quotations from Davis in this chapter are from this interview.

53. Kerstin Carr and Thea Walsh, interview by Ben Green, April 12, 2017. All quotations from Carr and Walsh in this chapter are from this interview.

54. Calthorpe Associates et al., "insight2050 Scenario Results Report" (February 26, 2015), p. 6, http://getinsight2050.org/wp-content/uploads/2015/03/2015_02_26 -insight2050-Report.pdf.

55. Calthorpe Associates et al., "insight2050 Scenario Results Report," 18–19.

56. Carla Bailo, interview by Ben Green, May 9, 2017. All quotations from Bailo in this chapter are from this interview.

57. City of Columbus, "Columbus Smart City Application" (2016), p. 6, https:// www.transportation.gov/sites/dot.gov/files/docs/Columbus OH Vision Narrative.pdf.

58. City of Columbus, "Linden Infant Mortality Profile" (2018), http://celebrateone .info/wp-content/uploads/2018/03/Linden_IMProfile_9.7.pdf.

59. Smart Columbus, "Smart Columbus Connects Linden Meeting Summary" (2017).

60. Laura Bliss, "Columbus Now Says 'Smart' Rides for Vulnerable Moms Are Coming," *CityLab* (December 1, 2017), https://www.citylab.com/transportation/2017/12/columbus-now-says-smart-rides-for-vulnerable-moms-are-coming/547013/.

61. Laura Bliss, "Who Wins When a City Gets Smart?," *CityLab* (November 1, 2017), https://www.citylab.com/transportation/2017/11/when-a-smart-city-doesnt-have-all-the-answers/542976/.

62. "Buggy Capital of the World," *Columbus Dispatch*, Blog (July 29, 2015), http://www.dispatch.com/content/blogs/a-look-back/2015/07/buggy-capital-of-the-world.html.

63. Thomas J. Misa, "Controversy and Closure in Technological Change: Constructing 'Steel,'" in *Shaping Technology / Building Society: Studies in Sociotechnical Change*, ed. Wiebe E. Bijker and John Law (Cambridge, MA: MIT Press, 1992), 110, 111.

64. Norton, *Fighting Traffic*, 2.

Chapter 3

1. Dominik Schiener, "Liquid Democracy: True Democracy for the 21st Century," Medium: Organizer Sandbox (November 23, 2015), https://medium.com/organizer-sandbox/liquid-democracy-true-democracy-for-the-21st-century-7c66f5e53b6f.

2. Gavin Newsom and Lisa Dickey, *Citizenville: How to Take the Town Square Digital and Reinvent Government* (New York: Penguin, 2014), 13, 10.

3. Mark Zuckerberg, "Facebook's Letter from Mark Zuckerberg—Full Text," *The Guardian*, February 1, 2012, https://www.theguardian.com/technology/2012/feb/01/facebook-letter-mark-zuckerberg-text.

4. Sean Parker, quoted in Anthony Ha, "Sean Parker: Defeating SOPA Was the 'Nerd Spring,'" *TechCrunch*, March 12, 2012, https://techcrunch.com/2012/03/12/sean-parker-defeating-sopa-was-the-nerd-spring/.

5. Nathan Daschle, quoted in Steve Friess, "Son of Dem Royalty Creates a Ruck.us," *Politico*, June 26, 2012, http://www.politico.com/story/2012/06/son-of-democratic-party-royalty-creates-a-ruckus-077847.

6. "Brigade," https://www.brigade.com.

7. Ferenstein Wire, "Sean Parker Explains His Plans to 'Repair Democracy' with a New Social Network," *Fast Company* (2015), https://www.fastcompany.com/3047571/sean-parker-explains-his-plans-to-repair-democracy-with-a-new-social-network.

8. Kim-Mai Cutler and Josh Constine, "Sean Parker's Brigade App Enters Private Beta as a Dead-Simple Way of Taking Political Positions," *TechCrunch*, June 17, 2015, https://techcrunch.com/2015/06/17/sean-parker-brigade/.

9. "Textizen," https://www.textizen.com.

10. Christopher Smart, "What Do You Like, Don't Like?—Text It to Salt Lake City," *Salt Lake Tribune*, August 20, 2012, http://archive.sltrib.com/story.php?ref=/sltrib/news/54728901-78/lake-salt-text-plan.html.csp.

11. Chante Lantos-Swett, "Leveraging Technology to Improve Participation: Textizen and Oregon's Kitchen Table," *Challenges to Democracy*, Blog (April 4, 2016), http://www.challengestodemocracy.us/home/leveraging-technology-to-improve-participation-textizen-and-oregons-kitchen-table/.

12. Thomas M. Menino, "Inaugural Address" (January 4, 2010), p. 5, https://www.cityof boston.gov/TridionImages/2010%20Thomas%20M%20%20Menino%20Inaugural %20final_tcm1-4838.pdf.

13. Jimmy Daly, "10 Cities With 311 iPhone Applications," *StateTech*, August 10, 2012, https://statetechmagazine.com/article/2012/08/10-cities-311-iphone-applications.

14. IBM, "What Is a Self-Service Government?," *The Atlantic* [advertising], http://www.theatlantic.com/sponsored/ibm-transformation/what-is-a-self-service-govern ment/248/.

15. Alexis de Tocqueville, *Democracy in America*, ed. Max Lerner and J.-P. Mayer, trans. George Lawrence (New York: Harper and Row, 1966), 2:522.

16. Evan Halper, "Napster Co-founder Sean Parker Once Vowed to Shake Up Washington—So How's That Working Out?," *Los Angeles Times*, August 4, 2016, http://www.latimes.com/politics/la-na-pol-sean-parker-20160804-snap-story.html.

17. Sean Parker, quoted in Greg Ferenstein, "Brigade: New Social Network from Facebook Co-founder Aims to 'Repair Democracy,'" *The Guardian*, June 17, 2015, https://www.theguardian.com/media/2015/jun/17/brigade-social-network-voter -turnout-sean-parker.

18. Christopher Fry and Henry Lieberman, *Why Can't We All Just Get Along?* (self-published, 2018), 257, 266, https://www.whycantwe.org/.

19. Corey Robin, quoted in Emma Roller, "'Victory Can Be a Bit of a Bitch': Corey Robin on the Decline of American Conservatism," *Splinter*, September 1, 2017), https://splinter news.com/victory-can-be-a-bit-of-a-bitch-corey-robin-on-the-dec-1798679236.

20. Bruno Latour, *The Pasteurization of France*, trans. Alan Sheridan and John Law (Cambridge, MA: Harvard University Press, 1993), 210.

21. Archon Fung, Hollie Russon Gilman, and Jennifer Shkabatur, "Six Models for the Internet + politics," *International Studies Review* 15, no. 1 (2013): 33, 37, 42, 45.

22. Schiener, "Liquid Democracy."

23. Hahrie Han, *How Organizations Develop Activists: Civic Associations and Leadership in the 21st Century* (New York: Oxford University Press, 2014).

24. Han, *How Organizations Develop Activists*, 95.

25. Han, *How Organizations Develop Activists*, 140–141.

26. Zeynep Tufekci, *Twitter and Tear Gas: The Power and Fragility of Networked Protest* (New Haven: Yale University Press, 2017), 200–201.

27. Han, *How Organizations Develop Activists*, 153.

28. Ryan W. Buell, Ethan Porter, and Michael I. Norton, "Surfacing the Submerged State: Operational Transparency Increases Trust in and Engagement with Government," Harvard Business School Working Paper No. 14-034 (November 2013; rev. March 2018).

29. Daniel Tumminelli O'Brien et al., "Uncharted Territoriality in Coproduction: The Motivations for 311 Reporting," *Journal of Public Administration Research and Theory* 27, no. 2 (2017): 331.

30. Ariel White and Kris-Stella Trump, "The Promises and Pitfalls of 311 Data," *Urban Affairs Review* 54, no. 4 (2016): 794–823, https://doi.org/10.1177/1078087416673202.

31. Kay Lehman Schlozman, Sidney Verba, and Henry E. Brady, *The Unheavenly Chorus: Unequal Political Voice and the Broken Promise of American Democracy* (Princeton, NJ: Princeton University Press, 2012), 6, 8.

32. Nancy Burns, Kay Lehman Schlozman, and Sidney Verba, *The Private Roots of Public Action: Gender, Equality, and Political Participation* (Cambridge, MA: Harvard University Press, 2001), 360.

33. Monica C. Bell, "Police Reform and the Dismantling of Legal Estrangement," *Yale Law Journal* 126 (2017): 2054 (abstract), 2085, 2057, 2101.

34. Bell, "Police Reform and the Dismantling of Legal Estrangement," 2141.

35. Michael Lipsky, *Street-Level Bureaucracy: Dilemmas of the Individual in Public Services*, 30th anniversary ed. (New York: Russell Sage Foundation, 2010), 3.

36. Matthew Desmond, Andrew V. Papachristos, and David S. Kirk, "Police Violence and Citizen Crime Reporting in the Black Community," *American Sociological Review* 81, no. 5 (2016): 857–876, 857 (abstract).

37. Elizabeth S. Anderson, "What is the Point of Equality?" *Ethics* 109, no. 2 (1999): 313.

38. Catherine E. Needham, "Customer Care and the Public Service Ethos," *Public Administration* 84, no. 4 (2006): 857–858.

39. Catherine Needham, *Citizen-Consumers: New Labour's Marketplace Democracy* (London: Catalyst, 2003), 6.

40. Jane E. Fountain, "Paradoxes of Public Sector Customer Service," *Governance* 14, no. 1 (2001): 56.

41. Dietmar Offenhuber, "The Designer as Regulator: Design Patterns and Categorization in Citizen Feedback Systems" (paper delivered at the Workshop on Big Data and Urban Informatics, Chicago, August 2014).

42. Daniel Tumminelli O'Brien, Eric Gordon, and Jessica Baldwin, "Caring about the Community, Counteracting Disorder: 311 Reports of Public Issues as Expressions of Territoriality," *Journal of Environmental Psychology* 40 (2014): 324–325.

43. James J. Feigenbaum and Andrew Hall, "How High-Income Areas Receive More Service from Municipal Government: Evidence from City Administrative Data" (2015), https://ssrn.com/abstract=2631106.

44. Abdallah Fayyad, "The Criminalization of Gentrifying Neighborhoods," *The Atlantic*, December 20, 2017, https://www.theatlantic.com/politics/archive/2017/12/the-criminalization-of-gentrifying-neighborhoods/548837/.

45. Al Baker, J. David Goodman, and Benjamin Mueller, "Beyond the Chokehold: The Path to Eric Garner's Death," *New York Times*, June 13, 2015, https://www.nytimes.com/2015/06/14/nyregion/eric-garner-police-chokehold-staten-island.html.

46. Jathan Sadowski and Frank Pasquale, "The Spectrum of Control: A Social Theory of the Smart City," *First Monday* 20, no. 7 (2015), http://firstmonday.org/article/view/5903/4660; they quote Stephen Goldsmith and Susan Crawford, *The Responsive City: Engaging Communities through Data-Smart Governance* (San Francisco: Jossey-Bass, 2014), 4.

47. Virgina Eubanks, *Automating Inequality: How High-Tech Tools Profile, Police, and Punish the Poor* (New York: St. Martin's Press, 2018), 136–138.

48. Rhema Vaithianathan, "Big Data Should Shrink Bureaucracy Big Time," *Stuff*, October 18, 2016, https://www.stuff.co.nz/national/politics/opinion/85416929/rhema-vaithianathan-big-data-should-shrink-bureaucracy-big-time.

49. Adam Forrest, "Detroit Battles Blight through Crowdsourced Mapping Project," *Forbes*, June 22, 2015, https://www.forbes.com/sites/adamforrest/2015/06/22/detroit-battles-blight-through-crowdsourced-mapping-project.

50. NYC Mayor's Office of Operations, "Hurricane Sandy Response," *NYC Customer Service Newsletter* 5, no. 2 (February 2013), https://www1.nyc.gov/assets/operations/downloads/pdf/nyc_customer_service_newsletter_volume_5_issue_2.pdf.

51. Joshua Tauberer, "So You Want to Reform Democracy," Medium: Civic Tech Thoughts from JoshData (November 22, 2015), https://medium.com/civic-tech-thoughts-from-joshdata/so-you-want-to-reform-democracy-7f3b1ef10597.

52. Mitch Weiss, interview by Ben Green, May 16, 2017.

53. Steven Walter, interview by Ben Green, April 20, 2017. All quotations from Walter in this chapter are from this interview.

54. Marshall Berman, "Take It to the Streets: Conflict and Community in Public Space," *Dissent* 33, no. 4 (1986): 477.

55. Cyndi Lauper, "Girls Just Want to Have Fun (Official Video)" (1983), https://www.youtube.com/watch?v=PIb6AZdTr-A.

56. John Hughes, dir., *Ferris Bueller's Day Off* (Paramount Pictures, 1986); for the parade scene, see "Ferris Bueller's Parade" (1986), https://www.youtube.com/watch?v=tRcv4nokK50.

57. Berman, "Take It to the Streets," 478–479.

58. Eric Gordon and Jessica Baldwin-Philippi, "Playful Civic Learning: Enabling Lateral Trust and Reflection in Game-based Public Participation," *International Journal of Communication* 8 (2014): 759.

59. Eric Gordon, "Civic Technology and the Pursuit of Happiness," *Governing* (2016), http://www.governing.com/cityaccelerator/civic-technology-and-the-pursuit-of-happiness.html.

60. Eric Gordon, interview by Ben Green, May 9, 2017.

61. Gordon and Baldwin-Philippi, "Playful Civic Learning," 759 (abstract).

62. Detroit Future City, "2012 Detroit Strategic Framework Plan" (2012), p. 730, https://detroitfuturecity.com/wp-content/uploads/2014/12/DFC_Full_2nd.pdf. Community PlanIt was launched in Detroit under the name "Detroit 24/7 Outreach."

63. Gordon and Baldwin-Philippi, "Playful Civic Learning," 777, 772, 773.

64. Eric Gordon and Stephen Walter, "Meaningful Inefficiencies: Resisting the Logic of Technological Efficiency in the Design of Civic Systems," in *Civic Media: Technology, Design, Practice*, ed. Eric Gordon and Paul Mihailidis (Cambridge, MA: MIT Press, 2016), 254, 246.

65. Gordon and Walter, "Meaningful Inefficiencies," 244, 251.

66. Gordon, interview by Green.

67. Gordon and Walter, "Meaningful Inefficiencies," 263.

68. Hollie Russon Gilman, *Democracy Reinvented: Participatory Budgeting and Civic Innovation in America* (Washington, DC: Brookings Institution Press, 2016), 14.

69. Gilman, *Democracy Reinvented*, 74.

70. Gilman, *Democracy Reinvented*, 14.

71. Gilman, *Democracy Reinvented*, 90, 86, 115, 11.

72. Gilman, *Democracy Reinvented*, 87.

73. Rafael Sampaio and Tiago Peixoto, "Electronic Participatory Budgeting: False Dilemmas and True Complexities," in *Hope for Democracy: 25 Years of Participatory Budgeting Worldwide*, ed. Nelson Dias (São Brás de Alportel, Portugal: In Loco Association, 2014), 423.

74. Gilman, *Democracy Reinvented*, 11.

75. Alyssa Lane, interview by Ben Green, August 24, 2017. All quotations from Lane in this chapter are from this interview.

76. Lynn M. Sanders, "Against Deliberation," *Political Theory* 25, no. 3 (1997): 347–376.

77. Gilman, *Democracy Reinvented*, 13–14.

78. Gilman, *Democracy Reinvented*, 116, 11.

Chapter 4

1. Dan Wagner and Rich Sevieri, interview by Ben Green, March 2, 2017, Cambridge, MA. All quotations from Wagner and Sevieri in this chapter are from this interview.

2. Cynthia Rudin, interview by Ben Green, February 18, 2017.

3. Deborah Lamm Weisel, "Burglary of Single-Family Houses," U.S. Department of Justice, Office of Community Oriented Policing Services, Problem-Oriented Guides for Police Series No. 18 (2002), http://www.popcenter.org/problems/pdfs/burglary_of_single-family_houses.pdf.

4. Wagner and Sevieri, interview by Green.

5. Tong Wang et al., "Finding Patterns with a Rotten Core: Data Mining for Crime Series with Cores," *Big Data* 3, no. 1 (2015): 3–21, http://doi.org/10.1089/big.2014.0021.

6. Wang et al., "Finding Patterns with a Rotten Core," 16–17.

7. Chris Anderson, "The End of Theory: The Data Deluge Makes the Scientific Method Obsolete," *Wired*, June 23, 2008, https://www.wired.com/2008/06/pb-theory/.

8. Amir Efrati, "Uber Finds Deadly Accident Likely Caused by Software Set to Ignore Objects on Road," *The Information*, May 7, 2018, https://www.theinformation.com/ articles/uber-finds-deadly-accident-likely-caused-by-software-set-to-ignore-objects -on-road.

9. For African Americans vs. whites, see Marianne Bertrand and Sendhil Mullainathan, "Are Emily and Greg More Employable than Lakisha and Jamal? A Field Experiment on Labor Market Discrimination," *American Economic Review* 94, no. 4 (2004): 991–1013; for women vs. men, see Ernesto Reuben, Paola Sapienza, and Luigi Zingales, "How Stereotypes Impair Women's Careers in Science," *Proceedings of the National Academy of Sciences* 111, no. 12 (2014): 4403–4408.

10. Stella Lowry and Gordon Macpherson, "A Blot on the Profession," *British Medical Journal* 296, no. 6623 (1988): 657–658.

11. Jeffrey Dastin, "Amazon Scraps Secret AI Recruiting Tool that Showed Bias Against Women," *Reuters*, October 9, 2018, https://www.reuters.com/article/us-amazon -com-jobs-automation-insight/amazon-scraps-secret-ai-recruiting-tool-that-showed -bias-against-women-idUSKCN1MK08G.

12. David Robinson, interview by Ben Green, February 21, 2017. Unless otherwise specified, quotations from Robinson in this chapter are from this interview.

13. PredPol, "Proven Crime Reduction Results" [2018], http://www.predpol.com/ results/.

14. Andrew G. Ferguson, "The Allure of Big Data Policing," *PrawfsBlawg* (May 25, 2017), http://prawfsblawg.blogs.com/prawfsblawg/2017/05/the-allure-of-big-data-policing.html.

15. "A brilliantly smart idea": Gillian Tett, "Mapping Crime—or Stirring Hate?," *Financial Times*, August 22, 2014, https://www.ft.com/content/200bebee-28b9-11e4 -8bda-00144feabdc0; "stop crime before it starts"; Joel Rubin, "Stopping Crime before It Starts," *Los Angeles Times*, August 21, 2010, http://articles.latimes.com/2010/ aug/21/local/la-me-predictcrime-20100427-1. The interview was with Zach Friend on *The War Room*, hosted by Jennifer Granholm, Current TV, January 16, 2013, posted as "PredPol on Current TV with Santa Cruz Crime Analyst Zach Friend" (2013), https://www.youtube.com/watch?v=8uKor0nfsdQ. For his involvement with Pred-Pol, see Darwin Bond-Graham, "All Tomorrow's Crimes: The Future of Policing Looks a Lot Like Good Branding," *SF Weekly*, October 30, 2013, http://www.sfweekly.com /news/all-tomorrows-crimes-the-future-of-policing-looks-a-lot-like-good-branding/.

16. David Robinson and Logan Koepke, "Stuck in a Pattern," *Upturn* (2016), https:// www.teamupturn.org/reports/2016/stuck-in-a-pattern/.

17. Tim Cushing, "'Predictive Policing' Company Uses Bad Stats, Contractually-Obligated Shills to Tout Unproven 'Successes,'" *Techdirt*, November 1, 2013, https:// www.techdirt.com/articles/20131031/13033125091/predictive-policing-company -uses-bad-stats-contractually-obligated-shills-to-tout-unproven-successes.shtml.

18. Philip Stark, chair of the statistics department at UC Berkeley, quoted in Bond-Graham, "All Tomorrow's Crimes."

19. John Hollywood, quoted in Mara Hvistendahl, "Can 'Predictive Policing' Prevent Crime Before It Happens?," *Science*, September 28, 2016, http://www.sciencemag .org/news/2016/09/can-predictive-policing-prevent-crime-it-happens.

20. Priscillia Hunt, Jessica Saunders, and John S. Hollywood, *Evaluation of the Shreveport Predictive Policing Experiment*, RR-531-NIJ (Santa Monica, CA: RAND Corporation, 2014), 33.

21. Brett Goldstein, quoted in Tett, "Mapping Crime—or Stirring Hate?"

22. Sean Malinowski, quoted in Justin Jouvenal, "Police Are Using Software to Predict Crime. Is It a 'Holy Grail' or Biased against Minorities?," *Washington Post*, November 17, 2016, https://www.washingtonpost.com/local/public-safety/police-are-using -software-to-predict-crime-is-it-a-holy-grail-or-biased-against-minorities/2016/11/17/ 525a6649-0472-440a-aae1-b283aa8e5de8_story.html.

23. Darrin Lipsomb, quoted in Jack Smith, "'Minority Report' Is Real—And It's Really Reporting Minorities," *Mic*, November 9, 2015, https://mic.com/articles/127739/ minority-reports-predictive-policing-technology-is-really-reporting-minorities.

24. Carl B. Klockars, "Some Really Cheap Ways of Measuring What Really Matters," in *Measuring What Matters: Proceedings from the Policing Research Institute Meetings* (Washington, DC: National Institute of Justice, 1999), 191.

25. See Michelle Alexander, *The New Jim Crow: Mass Incarceration in the Age of Colorblindness* (New York: New Press, 2012).

26. See Peter Moskos, *Cop in the Hood: My Year Policing Baltimore's Eastern District* (Princeton, NJ: Princeton University Press, 2008).

27. See Jeffrey Reiman and Paul Leighton, *The Rich Get Richer and the Poor Get Prison: Ideology, Class, and Criminal Justice* (New York: Routledge, 2015).

28. Sam Lavigne, Brian Clifton, and Francis Tseng, "Predicting Financial Crime: Augmenting the Predictive Policing Arsenal," *The New Inquiry* (2017), https://whitecollar .thenewinquiry.com/static/whitepaper.pdf.

29. See Paul Butler, *Chokehold: Policing Black Men* (New York: New Press, 2017).

30. Jacob Metcalf, "Ethics Review for Pernicious Feedback Loops," Medium: Data & Society: Points (November 7, 2016), https://points.datasociety.net/ethics-review-for -pernicious-feedback-loops-9a7ede4b610e.

31. Kristian Lum and William Isaac, "To Predict and Serve?," *Significance* 13, no. 5 (2016): 17.

32. Kristian Lum, "Predictive Policing Reinforces Police Bias," Human Rights Data Analysis Group (2016), https://hrdag.org/2016/10/10/predictive-policing-reinforces -police-bias/.

33. Jeremy Heffner, interview by Ben Green, March 18, 2017. All quotations from Heffner in this chapter are from this interview.

34. HunchLab, "Next Generation Predictive Policing," https://www.hunchlab.com; Amaury Murgado, "Developing a Warrior Mindset," *POLICE Magazine*, May 24, 2012, http://www.policemag.com/channel/patrol/articles/2012/05/warrior-mindset.aspx.

35. Hunt, Saunders, and Hollywood, "Evaluation of the Shreveport Predictive Policing Experiment," 12.

36. Nick O'Malley, "To Predict and to Serve: The Future of Law Enforcement," *Sydney Morning Herald*, March 30, 2013, http://www.smh.com.au/world/to-predict-and-to -serve-the-future-of-law-enforcement-20130330-2h0rb.html.

37. Sharad Goel, Justin M. Rao, and Ravi Shroff, "Precinct or Prejudice? Understanding Racial Disparities in New York City's Stop-and-Frisk Policy," *Annals of Applied Statistics* 10, no. 1 (2016): 365–394.

38. Ben Green, Thibaut Horel, and Andrew V. Papachristos, "Modeling Contagion through Social Networks to Explain and Predict Gunshot Violence in Chicago, 2006 to 2014," *JAMA Internal Medicine* 177, no. 3 (2017): 326–333, https://doi.org/10.1001/ jamainternmed.2016.8245.

39. David H. Bayley, *Police for the Future* (New York: Oxford University Press, 1996), 3.

40. Christopher M. Sullivan and Zachary P. O'Keeffe, "Evidence That Curtailing Proactive Policing Can Reduce Major Crime," *Nature Human Behaviour* 1 (2017): 735, 730 (title).

41. John Chasnoff, quoted in Maurice Chammah, "Policing the Future," *The Verge* (2016), http://www.theverge.com/2016/2/3/10895804/st-louis-police-hunchlab -predictive-policing-marshall-project.

42. Robert Sullivan, interview by Ben Green, March 21, 2017. All quotations from Robert Sullivan in this chapter are from this interview.

43. National Association of Counties, "Mental Health and Criminal Justice Case Study: Johnson County, Kan." (2015), http://www.naco.org/sites/default/files/ documents/Johnson%20County%20Mental%20Health%20and%20Jails%20 Case%20Study_FINAL.pdf; "Nine Additional Cities Join Johnson County's Co-responder Program," press release, Johnson County, Kansas, July 18, 2016, https://jocogov.org/press-release/nine-additional-cities-join-johnson-county's-co -responder-program.

44. Sullivan, interview by Green.

45. The Data-Driven Justice Initiative, "Data-Driven Justice Playbook" (2016), p. 3, http:// www.naco.org/sites/default/files/documents/DDJ%20Playbook%20Discussion %20Draft%2012.8.16_1.pdf.

46. The Data-Driven Justice Initiative, "Data-Driven Justice Playbook."

47. Lynn Overmann, "Launching the Data-Driven Justice Initiative: Disrupting the Cycle of Incarceration," *The Obama White House* (2016), https://medium.com/@Obama WhiteHouse/launching-the-data-driven-justice-initiative-disrupting-the-cycle-of -incarceration-e222448a64cf.

48. Peter Koutoujian, quoted in "Middlesex Police Discuss Data-Driven Justice Initiative," *Wicked Local Arlington*, December 30, 2016, http://arlington.wickedlocal. com/news/20161230/middlesex-police-discuss-data-driven-justice-initiative.

49. Overmann, "Launching the Data-Driven Justice Initiative."

50. Thomas E. Perez, "Investigation of the Miami-Dade County Jail," U.S. Department of Justice, Civil Rights Division (August 24, 2011), p. 10, https://www.clearinghouse .net/chDocs/public/JC-FL-0021-0004.pdf.

51. Overmann, "Launching the Data-Driven Justice Initiative."

52. The White House Office of the Press Secretary, "FACT SHEET: Launching the Data-Driven Justice Initiative: Disrupting the Cycle of Incarceration" (June 30, 2016), https://obamawhitehouse.archives.gov/the-press-office/2016/06/30/fact-sheet -launching-data-driven-justice-initiative-disrupting-cycle.

53. Overmann, "Launching the Data-Driven Justice Initiative."

54. Will Engelhardt et al., "Sharing Information between Behavioral Health and Criminal Justice Systems," Council of State Governments Justice Center (March 31, 2016), p. 3, https://csgjusticecenter.org/wp-content/uploads/2016/03/JMHCP-Info-Sharing -Webinar.pdf.

55. Matthew J. Bauman et al., "Reducing Incarceration through Prioritized Interventions," in *COMPASS '18: Proceedings of the 1st ACM SIGCAS Conference on Computing and Sustainable Societies* (2018).

56. Bauman et al., "Reducing Incarceration through Prioritized Interventions," 7; Center for Data Science and Public Policy, University of Chicago, "Data-Driven Justice Initiative: Identifying Frequent Users of Multiple Public Systems for More Effective Early Assistance" (2018), https://dsapp.uchicago.edu/projects/criminal-justice/ data-driven-justice-initiative/.

57. Steve Yoder, interview by Ben Green, March 27, 2017. All quotations from Yoder in the chapter are from this interview.

58. The Laura and John Arnold Foundation, "Laura and John Arnold Foundation to continue data-driven criminal justice effort launched under the Obama Administration," press release, January 23, 2017, http://www.arnoldfoundation. org/laura-john-arnold-foundation-continue-data-driven-criminal-justice-effort- launched-obama-administration/.

59. National Association of Counties, "Data-Driven Justice: Disrupting the Cycle of Incarceration" [2015], http://www.naco.org/resources/signature-projects/data-driven -justice.

60. PredPol, "How Predictive Policing Works" [2018], http://www.predpol.com/ how-predictive-policing-works/.

61. Richard Berk, in Craig Atkinson, dir., *Do Not Resist* (Vanish Films, 2016).

62. Dominic Griffin, "'Do Not Resist' Traces the Militarization of Police with Unprecedented Access to Raids and Unrest," *Baltimore City Paper*, November 1, 2016, http://www.citypaper.com/film/film/bcp-110216-screens-do-not-resist-20161101 -story.html.

63. Joshua Brustein, "This Guy Trains Computers to Find Future Criminals," *Bloomberg* (2016), https://www.bloomberg.com/features/2016-richard-berk-future-crime/.

64. Thomas P. Bonczar, "Prevalence of Imprisonment in the U.S. Population, 1974– 2001," *Bureau of Justice Statistics Special Report* (August 2003), p. 1, https://www.bjs .gov/content/pub/pdf/piusp01.pdf.

65. On government programs, see Richard Rothstein, *The Color of Law: A Forgotten History of How Our Government Segregated America* (New York: Liveright, 2017); on the war on drugs, see Alexander, *The New Jim Crow*.

66. IBM, "Predictive Analytics: Police Use Analytics to Reduce Crime" (2012), https://www.youtube.com/watch?v=iY3WRvXVogo.

67. Alex S. Vitale, *The End of Policing* (London: Verso, 2017), cover, 28.

68. Andrew V. Papachristos and Christopher Wildeman, "Network Exposure and Homicide Victimization in an African American Community," *American Journal of Public Health* 104, no. 1 (2014): 143–150.

69. Jeremy Gorner, "With Violence Up, Chicago Police Focus on a List of Likeliest to Kill, Be Killed," *Chicago Tribune*, July, 22, 2016, http://www.chicagotribune.com/news/ct-chicago-police-violence-strategy-met-20160722-story.html.

70. Jessica Saunders, Priscillia Hunt, and John S. Hollywood, "Predictions Put into Practice: A Quasi-Experimental Evaluation of Chicago's Predictive Policing Pilot," *Journal of Experimental Criminology* 12, no. 3 (2016): 366, 355.

71. Andrew V. Papachristos, "CPD's Crucial Choice: Treat Its List as Offenders or as Potential Victims?," *Chicago Tribune*, July 29, 2016, http://www.chicagotribune.com/news/opinion/commentary/ct-gun-violence-list-chicago-police-murder-perspec-0801-jm-20160729-story.html.

72. Ferguson, "The Allure of Big Data Policing."

Chapter 5

1. Langdon Winner, *The Whale and the Reactor: A Search for Limits in an Age of High Technology* (Chicago: University of Chicago Press, 1986), 55, 49, 52.

2. Cecilia Kang, "Unemployed Detroit Residents Are Trapped by a Digital Divide," *New York Times*, May 23, 2016, https://www.nytimes.com/2016/05/23/technology/unemployed-detroit-residents-are-trapped-by-a-digital-divide.html.

3. Letitia James and Ben Kallos, "New York City Digital Divide Fact Sheet," press release, March 16, 2017.

4. LinkNYC, "Find a Link," https://www.link.nyc/find-a-link.html.

5. New York City Office of the Mayor, "Mayor de Blasio Announces Public Launch of LinkNYC Program, Largest and Fastest Free Municipal Wi-Fi Network in the World" (February 18, 2016), http://www1.nyc.gov/office-of-the-mayor/news/184-16/mayor-de-blasio-public-launch-linknyc-program-largest-fastest-free-municipal#/0.

6. Dan Doctoroff, quoted in Nick Pinto, "Google Is Transforming NYC's Payphones into a 'Personalized Propaganda Engine,'" *Village Voice*, July 6, 2016, https://www.villagevoice.com/2016/07/06/google-is-transforming-nycs-payphones-into-a-personalized-propaganda-engine/.

7. LinkNYC, "Privacy Policy" (March 17, 2017), https://www.link.nyc/privacy-policy.html.

8. LinkNYC, "Privacy Policy."

9. Paul M. Schwartz and Daniel J. Solove, "The PII Problem: Privacy and a New Concept of Personally Identifiable Information," *NYU Law Review* 86 (2011): 1814–1895.

10. On phone location traces, see De Montjoye et al., "Unique in the Crowd"; on credit card transactions, see Yves-Alexandre de Montjoye et al., "Unique in the Shopping Mall: On the Reidentifiability of Credit Card Metadata," *Science* 347, no. 6221 (2015): 536–539.

11. Erica Klarreich, "Privacy by the Numbers: A New Approach to Safeguarding Data," *Quanta Magazine*, December 10, 2012, https://www.quantamagazine.org/a-mathematical-approach-to-safeguarding-private-data-20121210/.

12. Latanya Sweeney, "Simple Demographics Often Identify People Uniquely" (Carnegie Mellon University, Data Privacy Working Paper 3, 2000).

13. Anthony Tockar, "Riding with the Stars: Passenger Privacy in the NYC Taxicab Dataset," *Neustar Research*, September 15, 2014, https://research.neustar.biz/2014/09/15/riding-with-the-stars-passenger-privacy-in-the-nyc-taxicab-dataset/.

14. James Siddle, "I Know Where You Were Last Summer: London's Public Bike Data Is Telling Everyone Where You've Been," *The Variable Tree*, April 10, 2014, https://vartree.blogspot.co.uk/2014/04/i-know-where-you-were-last-summer.html.

15. On whom you know, see Nathan Eagle, Alex Sandy Pentland, and David Lazer, "Inferring Friendship Network Structure by Using Mobile Phone Data," *Proceedings of the National Academy of Sciences* 106, no. 36 (2009): 15274–15278; on where you will go next, see Lars Backstrom, Eric Sun, and Cameron Marlow, "Find Me If You Can: Improving Geographical Prediction with Social and Spatial Proximity" (paper presented at the Proceedings of the 19th International Conference on World Wide Web, Raleigh, NC, April 2010).

16. Andrew G. Reece and Christopher M. Danforth, "Instagram Photos Reveal Predictive Markers of Depression," *EPJ Data Science* 6, no. 15 (2017), https://doi.org/10.1140/epjds/s13688-017-0110-z.

17. Michal Kosinski, David Stillwell, and Thore Graepel, "Private Traits and Attributes Are Predictable from Digital Records of Human Behavior," *Proceedings of the National Academy of Sciences* 110, no. 15 (2013): 5802–5805.

18. Eben Moglen, quoted in Pinto, "Google Is Transforming NYC's Payphones."

19. Pinto, "Google Is Transforming NYC's Payphones."

20. Donna Lieberman, quoted in New York Civil Liberties Union, "City's Public Wi-Fi Raises Privacy Concerns" (March 16, 2016), https://www.nyclu.org/en/press-releases/nyclu-citys-public-wi-fi-raises-privacy-concerns.

21. Daniel J. Solove, *The Digital Person: Technology and Privacy in the Information Age* (New York: NYU Press, 2004).

22. DeRay McKesson, quoted in Jessica Guynn, "ACLU: Police Used Twitter, Facebook to Track Protests," *USA Today*, October 12, 2016, https://www.usatoday.com/story/tech/news/2016/10/11/aclu-police-used-twitter-facebook-data-track-protesters-baltimore-ferguson/91897034/.

23. Dia Kayyali, "The History of Surveillance and the Black Community," *Electronic Frontier Foundation* (February 13, 2014), https://www.eff.org/deeplinks/2014/02/history-surveillance-and-black-community.

24. On federal officials, see George Joseph, "Exclusive: Feds Regularly Monitored Black Lives Matter Since Ferguson," *The Intercept*, July 24, 2015, https://theintercept .com/2015/07/24/documents-show-department-homeland-security-monitoring -black-lives-matter-since-ferguson/; on local officials, see Nicole Ozer, "Police Use of Social Media Surveillance Software Is Escalating, and Activists Are in the Digital Crosshairs," Medium: ACLU of Northern CA (2016), https://medium.com/@ ACLU_NorCal/police-use-of-social-media-surveillance-software-is-escalating-and -activists-are-in-the-digital-d29d8f89c48.

25. Solove, *The Digital Person*, 34.

26. Solove, *The Digital Person*, 38, 37.

27. On Facebook and mood, see Adam D. I. Kramer, Jamie E. Guillory, and Jeffrey T. Hancock, "Experimental Evidence of Massive-Scale Emotional Contagion through Social Networks," *Proceedings of the National Academy of Sciences* 111, no. 24 (2014): 8788–8790; on Facebook and voting, see Robert M. Bond et al., "A 61-Million-Person Experiment in Social Influence and Political Mobilization," *Nature* 489, no. 7415 (2012): 295–298.

28. Christian Rudder, "We Experiment On Human Beings!," *The OkCupid Blog* (2014), https://theblog.okcupid.com/we-experiment-on-human-beings-5dd9fe280cd5.

29. Casey Johnston, "Denied for That Loan? Soon You May Thank Online Data Collection," *ArsTechnica* (2013), https://arstechnica.com/business/2013/10/denied-for -that-loan-soon-you-may-thank-online-data-collection/.

30. Mary Madden et al., "Privacy, Poverty and Big Data: A Matrix of Vulnerabilities for Poor Americans," *Washington University Law Review* 95 (2017): 53–125.

31. Virginia Eubanks, "Technologies of Citizenship: Surveillance and Political Learning in the Welfare System," in *Surveillance and Security: Technological Politics and Power in Everyday Life*, ed. Torin Monahan (New York: Routledge, 2006), 91.

32. John Gilliom, *Overseers of the Poor: Surveillance, Resistance, and the Limits of Privacy* (Chicago: University of Chicago Press, 2001), 6, 129, 1.

33. John Podesta et al., *Big Data: Seizing Opportunities, Preserving Values* (Washington, DC: Executive Office of the President, 2014).

34. "The Rise of Workplace Spying," *The Week*, July 8, 2015, http://theweek.com/ articles/564263/rise-workplace-spying.

35. Federal Trade Commission, *Data Brokers: A Call for Transparency and Accountability* (Washington, DC: Federal Trade Commission, 2014).

36. Virgina Eubanks, *Automating Inequality: How High-Tech Tools Profile, Police, and Punish the Poor* (New York: St. Martin's Press, 2018).

37. On "three million," see Jason Henry, "Los Angeles Police, Sheriff's Scan over 3 Million License Plates A Week," *San Gabriel Valley Tribune*, August 26, 2014, https:// www.sgvtribune.com/2014/08/26/los-angeles-police-sheriffs-scan-over-3-million -license-plates-a-week/; on ICE, see April Glaser, "Sanctuary Cities Are Handing ICE a Map," *Slate*, March 13, 2018, https://slate.com/technology/2018/03/how-ice-may -be-able-to-access-license-plate-data-from-sanctuary-cities-and-use-it-for-arrests.html.

38. The Leadership Conference on Civil and Human Rights and Upturn, "Police Body Worn Cameras: A Policy Scorecard" (November 2017), https://www.bwcscorecard.org.

39. Ava Kofman, "Real-Time Face Recognition Threatens to Turn Cops' Body Cameras into Surveillance Machines," *The Intercept*, March 22, 2017, https://theintercept.com/2017/03/22/real-time-face-recognition-threatens-to-turn-cops-body-cameras-into-surveillance-machines/.

40. Martin Kaste, "Orlando Police Testing Amazon's Real-Time Facial Recognition," *National Public Radio*, May 22, 2018, https://www.npr.org/2018/05/22/613115969/orlando-police-testing-amazons-real-time-facial-recognition.

41. See Federal Trade Commission, *Data Brokers: A Call for Transparency and Accountability.*

42. Jonas Lerman, "Big Data and Its Exclusions," *Stanford Law Review* 66 (2013): 55–63.

43. Madden et al., "Privacy, Poverty and Big Data."

44. Ross Garlick, "Privacy Inequality Is Coming, and It Does Not Look Pretty," *Fordham Political Review*, March 17, 2015, http://fordhampoliticalreview.org/privacy-inequality-is-coming-and-it-does-not-look-pretty/.

45. City and County of San Francisco, "Apps: Transportation," *San Francisco Data* [2018], http://apps.sfgov.org/showcase/apps-categories/transportation/.

46. City of Philadelphia, "Open Budget," http://www.phila.gov/openbudget/.

47. On sexual assault victims, see Andrea Peterson, "Why the Names of Six People Who Complained of Sexual Assault Were Published Online by Dallas Police," *Washington Post*, April 29, 2016, https://www.washingtonpost.com/news/the-switch/wp/2016/04/29/why-the-names-of-six-people-who-complained-of-sexual-assault-were-published-online-by-dallas-police/; on those carrying large sums of cash, see Claudia Vargas, "City Settles Gun Permit Posting Suit," *Philadelphia Inquirer*, July 23, 2014, http://www.philly.com/philly/news/local/20140723_City_settles_gun_permit_suit_for__1_4_million.html.

48. On medical information, see Klarreich, "Privacy by the Numbers"; on political affiliation, see Ethan Chiel, "Why the D.C. Government Just Publicly Posted Every D.C. Voter's Address Online," *Splinter*, June 14, 2016, https://splinternews.com/why-the-d-c-government-just-publicly-posted-every-d-c-1793857534.

49. Ben Green et al., "Open Data Privacy: A Risk-Benefit, Process-Oriented Approach to Sharing and Protecting Municipal Data," *Berkman Klein Center for Internet & Society Research Publication* (2017), http://nrs.harvard.edu/urn-3:HUL.InstRepos:30340010.

50. Green et al., "Open Data Privacy," 58–61.

51. Phil Diehl, "Malware Blamed for City's Data Breach," *San Diego Tribune*, September 12, 2017, http://www.sandiegouniontribune.com/communities/north-county/sd-no-malware-letter-20170912-story.html.

52. Selena Larson, "Uber's Massive Hack: What We Know," *CNN*, November 22, 2017, http://money.cnn.com/2017/11/22/technology/uber-hack-consequences-cover-up/index.html.

53. Bruce Schneier, *Click Here to Kill Everybody: Security and Survival in a Hyper-connected World* (New York: W. W. Norton, 2018).

54. Bruce Schneier, "Data Is a Toxic Asset, So Why Not Throw It Out?," *CNN*, March 1, 2016, http://www.cnn.com/2016/03/01/opinions/data-is-a-toxic-asset-opinion-schneier/index.html.

55. Pinto, "Google Is Transforming NYC's Payphones."

56. Douglas Rushkoff, quoted in Pinto, "Google Is Transforming NYC's Payphones."

57. On assigning students to schools, see Alvin Roth, "Why New York City's High School Admissions Process Only Works Most of the Time," *Chalkbeat*, July 2, 2015, https://www.chalkbeat.org/posts/ny/2015/07/02/why-new-york-citys-high-school-admissions-process-only-works-most-of-the-time/; on evaluating teachers, see Cathy O'Neil, "Don't Grade Teachers with a Bad Algorithm," *Bloomberg*, May 15, 2017, https://www.bloomberg.com/view/articles/2017-05-15/don-t-grade-teachers-with-a-bad-algorithm; on detecting Medicaid fraud, see Natasha Singer, "Bringing Big Data to the Fight against Benefits Fraud," *New York Times*, February 22, 2015, https://www.nytimes.com/2015/02/22/technology/bringing-big-data-to-the-fight-against-benefits-fraud.html; and on preventing fires, see Bob Sorokanich, "New York City Is Using Data Mining to Fight Fires," *Gizmodo* (2014), https://gizmodo.com/new-york-city-is-fighting-fires-with-data-mining-1509004543.

58. Jeff Asher and Rob Arthur, "Inside the Algorithm That Tries to Predict Gun Violence in Chicago," *New York Times*, June 13, 2017, https://www.nytimes.com/2017/06/13/upshot/what-an-algorithm-reveals-about-life-on-chicagos-high-risk-list.html.

59. Jeremy Gorner, "With Violence Up, Chicago Police Focus on a List of Likeliest to Kill, Be Killed," *Chicago Tribune*, July 22, 2016, http://www.chicagotribune.com/news/ct-chicago-police-violence-strategy-met-20160722-story.html.

60. On nondisclosure agreements, see Elizabeth E. Joh, "The Undue Influence of Surveillance Technology Companies on Policing," *New York University Law Review* 92 (2017): 101–130; on trade secrecy, see Rebecca Wexler, "Life, Liberty, and Trade Secrets: Intellectual Property in the Criminal Justice System," *Stanford Law Review* 70 (2018): 1343–1429.

61. On Intrado, see Justin Jouvenal, "The New Way Police Are Surveilling You: Calculating Your Threat 'Score,'" *Washington Post*, January 10, 2016, https://www.washingtonpost.com/local/public-safety/the-new-way-police-are-surveilling-you-calculating-your-threat-score/2016/01/10/e42bccac-8e15-11e5-baf4-bdf37355da0c_story.html; on Northpointe, see Frank Pasquale, "Secret Algorithms Threaten the Rule of Law," *MIT Technology Review*, June 1, 2017, https://www.technologyreview.com/s/608011/secret-algorithms-threaten-the-rule-of-law/.

62. Robert Brauneis and Ellen P. Goodman, "Algorithmic Transparency for the Smart City," *Yale Journal of Law and Technology* 20 (2018): 146–147.

63. *State v. Loomis*, 881 Wis. N.W.2d 749, 767 (2016).

64. Bernard E. Harcourt, "Risk as a Proxy for Race: The Dangers of Risk Assessment," *Federal Sentencing Reporter* 27, no. 4 (2015): 237–243.

65. Julia Angwin et al., "Machine Bias," *ProPublica*, May 23, 2016, https://www
.propublica.org/article/machine-bias-risk-assessments-in-criminal-sentencing.

66. Jon Kleinberg, Sendhil Mullainathan, and Manish Raghavan, "Inherent Trade-
Offs in the Fair Determination of Risk Scores," *arXiv.org* (2016), https://arxiv.org/
abs/1609.05807.

67. The New York City Council, "Int 1696-2017: Automated Decision Systems
Used by Agencies" (2017), http://legistar.council.nyc.gov/LegislationDetail.aspx
?ID=3137815&GUID=437A6A6D-62E1-47E2-9C42-461253F9C6D0.

68. The New York City Council, "Transcript of the Minutes of the Committee on
Technology," October 16, 2017, pp. 8–9, http://legistar.council.nyc.gov/View.ashx
?M=F&ID=5522569&GUID=DFECA4F2-E157-42AB-B598-BA3A8185E3FF.

69. The New York City Council, "Transcript of the Minutes of the Committee on
Technology," 7–8.

70. The New York City Council, "Int 1696-2017: Automated Decision Systems Used
by Agencies."

71. Julia Powles, "New York City's Bold, Flawed Attempt to Make Algorithms
Accountable," *New Yorker*, December 20, 2017, https://www.newyorker.com/tech/
elements/new-york-citys-bold-flawed-attempt-to-make-algorithms-accountable.

72. Array of Things, "Array of Things" (2016), http://arrayofthings.github.io.

73. Matt McFarland, "Chicago Gets Serious about Tracking Air Quality and Traffic Data," *CNN*, August 29, 2016, http://money.cnn.com/2016/08/29/technology/
chicago-sensors-data/index.html.

74. Denise Linn and Glynis Startz, "Array of Things Civic Engagement Report"
(August 2016), https://arrayofthings.github.io/engagement-report.html.

75. Array of Things, "Responses to Public Feedback" (2016), https://arrayofthings
.github.io/policy-responses.html.

76. Green et al., "Open Data Privacy," 34, 41.

77. Array of Things "Array of Things Operating Policies" (August 15, 2016), https://
arrayofthings.github.io/final-policies.html.

78. Brendan Kiley and Matt Fikse-Verkerk, "You Are a Rogue Device," *The Stranger*,
November 6, 2013, http://www.thestranger.com/seattle/you-are-a-rogue-device/
Content?oid=18143845.

79. Green et al., "Open Data Privacy," 89.

80. Michael Mattmiller, interview by Ben Green, August 3, 2017. All quotations
from Mattmiller in this chapter are from this interview.

81. City of Seattle, "City of Seattle Privacy Principles" (2015), https://www.seattle
.gov/Documents/Departments/InformationTechnology/City-of-Seattle-Privacy
-Principles-FINAL.pdf.

82. City of Seattle, "About the Privacy Program" (2018), http://www.seattle.gov/
tech/initiatives/privacy/about-the-privacy-program.

83. Rosalind Brazel, "City of Seattle Hires Ginger Armbruster as Chief Privacy
Officer," *Tech Talk Blog* (July 11, 2017), http://techtalk.seattle.gov/2017/07/11/city
-of-seattle-hires-ginger-armbruster-as-chief-privacy-officer/.

84. See Seattle Information Technology, "About the Surveillance Ordinance" (2018), https://www.seattle.gov/tech/initiatives/privacy/surveillance-technologies/about-surveillance-ordinance.

85. Ansel Herz, "How the Seattle Police Secretly—and Illegally—Purchased a Tool for Tracking Your Social Media Posts," *The Stranger*, September 28, 2016, https://www.thestranger.com/news/2016/09/28/24585899/how-the-seattle-police-secretlyand-illegallypurchased-a-tool-for-tracking-your-social-media-posts.

86. Ali Winston, "Palantir Has Secretly Been Using New Orleans to Test Its Predictive Policing Technology," *The Verge*, February 27, 2018, https://www.theverge.com/2018/2/27/17054740/palantir-predictive-policing-tool-new-orleans-nopd.

87. On Hattiesburg, see Haskel Burns, "Ordinance Looks at Police Surveillance Equipment," *Hattiesburg American*, October 28, 2016, https://www.hattiesburgamerican.com/story/news/local/hattiesburg/2016/10/28/ordinance-looks-police-surveillance-equipment/92899430/; on Oakland, see Ali Tadayon, "Oakland to Require Public Approval of Surveillance Tech," *East Bay Times*, May 2, 2018, https://www.eastbaytimes.com/2018/05/02/oakland-to-require-public-approval-of-surveillance-tech/; more generally, see American Civil Liberties Union, "Community Control over Police Surveillance" (2018), https://www.aclu.org/issues/privacy-technology/surveillance-technologies/community-control-over-police-surveillance.

88. Marc Groman, quoted in Jill R. Aitoro, "Defining Privacy Protection by Acknowledging What It's Not," *Federal Times*, March 8, 2016, http://www.federaltimes.com/story/government/interview/one-one/2016/03/08/defining-privacy-protection-acknowledging-what-s-not/81464556/.

89. Nigel Jacob, interview by Ben Green, April 7, 2017. All quotations from Jacob in this chapter are from this interview.

90. On open data, see Civic Analytics Network, "An Open Letter to the Open Data Community," Data-Smart City Solutions, March 3, 2017, https://datasmart.ash.harvard.edu/news/article/an-open-letter-to-the-open-data-community-988; on net neutrality, see Kimberly M. Aquilina, "50 US Cities Pen Letter to FCC Demanding Net Neutrality, Democracy," *Metro*, July 12, 2017, https://www.metro.us/news/local-news/net-neutrality-50-cities-letter-fcc-democracy.

91. Thomas Graham, "Barcelona Is Leading the Fightback against Smart City Surveillance," *Wired*, May 18, 2018, http://www.wired.co.uk/article/barcelona-decidim-ada-colau-francesca-bria-decode.

Chapter 6

1. "Sidewalk Labs," https://www.sidewalklabs.com.

2. Daniel L. Doctoroff, "Reimagining Cities from the Internet Up," Medium: Sidewalk Talk (November 30, 2016), https://medium.com/sidewalk-talk/reimagining-cities-from-the-internet-up-5923d6be63ba.

3. "Sidewalk Labs."

4. Amen Ra Mashariki, interview by Ben Green, May 24, 2017. All quotations from Mashariki in this chapter are from this interview, unless specified otherwise.

5. Allison T. Chamberlain, Jonathan D. Lehnert, and Ruth L. Berkelman, "The 2015 New York City Legionnaires' Disease Outbreak: A Case Study on a History-Making Outbreak," *Journal of Public Health Management and Practice* 23, no. 4 (2017): 414.

6. Mitsue Iwata, interview by Ben Green, July 25, 2017.

7. Joy Bonaguro, interview by Ben Green, August 9, 2017. All quotations from Bonaguro in this chapter are from this interview.

8. DataSF, "DataSF in Progress" (2018), https://datasf.org/progress/.

9. Bonaguro, interview by Green.

10. DataSF, "Data Quality" (2017), https://datasf.org/resources/data-quality/.

11. DataSF, "Data Standards Reference Handbook" (2018), https://datasf.gitbooks.io/draft-publishing-standards/.

12. DataSF, "DataScienceSF" (2017), https://datasf.org/science/.

13. DataSF, "Keeping Moms and Babies in Nutrition Program" (2018), https://datasf.org/showcase/datascience/keeping-moms-and-babies-in-nutrition-program/.

14. DataSF, "Eviction Alert System" (2018), https://datasf.org/showcase/datascience/eviction-alert-system/.

15. Mashariki, interview by Green.

16. Amen Ra Mashariki, "NYC Data Analytics" (presentation at Esri Senior Executive Summit, 2017), https://www.youtube.com/watch?v=ws8EQg5YlrY.

17. Seattle/King County Coalition on Homelessness, "2015 Street Count Results" (2015), http://www.homelessinfo.org/what_we_do/one_night_count/2015_results.php.

18. Daniel Beekman and Jack Broom, "Mayor, County Exec Declare 'State of Emergency' over Homelessness," *Seattle Times*, January 31, 2016, http://www.seattletimes.com/seattle-news/politics/mayor-county-exec-declare-state-of-emergency-over-homelessness/.

19. John Ryan, "After 10-Year Plan, Why Does Seattle Have More Homeless Than Ever?," *KUOW*, March 3, 2015, http://kuow.org/post/after-10-year-plan-why-does-seattle-have-more-homeless-ever.

20. Shakira Boldin, speaking in What Works Cities, "Tackling Homelessness in Seattle" (2017), https://www.youtube.com/watch?v=dzkblumT4XU.

21. City of Seattle, "Homelessness Investment Analysis" (2015), https://www.seattle.gov/Documents/Departments/HumanServices/Reports/HomelessInvestment Analysis.pdf.

22. Jason Johnson, interview by Ben Green, August 10, 2017. All quotations from Johnson in this chapter are from this interview.

23. Hanna Azemati and Christina Grover-Roybal, "Shaking Up the Routine: How Seattle Is Implementing Results-Driven Contracting Practices to Improve Outcomes for People Experiencing Homelessness," Harvard Kennedy School Government Performance Lab (September 2016), http://govlab.hks.harvard.edu/files/siblab/files/seattle_rdc_policy_brief_final.pdf.

24. Christina Grover-Roybal, interview by Ben Green, August 24, 2017. All quotations from Grover-Roybal in this chapter are from this interview.

25. Laura Melle, interview by Ben Green, April 12, 2017. All quotations from Melle in this chapter are from this interview.

26. Jeff Liebman, "Transforming the Culture of Procurement in State and Local Government," interview by Andy Feldman, *Gov Innovator* podcast (April 20, 2017), http://govinnovator.com/jeffrey_liebman_2017/.

27. "Results-Driven Contracting: An Overview," Harvard Kennedy School Government Performance Lab (2016), http://govlab.hks.harvard.edu/files/siblab/files/results-driven_contracting_an_overview_0.pdf.

28. Andrew Feldman and Jason Johnson, "How Better Procurement Can Drive Better Outcomes for Cities," *Governing*, October 12, 2017, http://www.governing.com/gov-institute/voices/col-cities-3-steps-procurement-reform-better-outcomes.html.

29. Azemati and Grover-Roybal, "Shaking Up the Routine."

30. Karissa Braxton, "City's Homeless Response Investments Are Housing More People," *City of Seattle Human Interests Blog* (May 31, 2018), http://humaninterests.seattle.gov/2018/05/31/citys-homeless-response-investments-are-housing-more-people/.

31. Boldin, speaking in What Works Cities, "Tackling Homelessness in Seattle."

32. Chris Anderson, "The End of Theory: The Data Deluge Makes the Scientific Method Obsolete," *Wired*, June 23, 2008, https://www.wired.com/2008/06/pb-theory/.

33. Tom Schenk, interview by Ben Green, August 8, 2017. All quotations from Schenk in this chapter are from this interview.

34. City of Chicago, "Food Inspection Forecasting" (2017), https://chicago.github.io/food-inspections-evaluation/.

35. Nigel Jacob, speaking in "Data-Driven Research, Policy, and Practice: Friday Opening Remarks" (Boston Area Research Initiative, March 10, 2017), https://www.youtube.com/watch?v=cRINlFFBHBo.

36. City of Boston, the Mayor's Office of New Urban Mechanics, "Civic Research Agenda" (May 15, 2018), https://www.boston.gov/departments/new-urban-mechanics/civic-research-agenda.

37. Kim Lucas, interview by Ben Green, May 6, 2017. All quotations from Lucas in this chapter are from this interview.

Chapter 7

1. Daniel L. Doctoroff, "Reimagining Cities from the Internet Up," Medium: Sidewalk Talk (November 30, 2016), https://medium.com/sidewalk-talk/reimagining-cities-from-the-internet-up-5923d6be63ba.

2. James C. Scott, *Seeing Like a State: How Certain Schemes to Improve the Human Condition Have Failed* (New Haven: Yale University Press, 1998), 11–22.

3. Scott, *Seeing Like a State*, 12.

4. Scott, *Seeing Like a State*, 21.

5. Scott, *Seeing Like a State*, 88.

6. Scott, *Seeing Like a State*, 4.

7. Ebenezer Howard, *Garden Cities of To-Morrow* (London: Swan Sonnenschein, 1902), 133.

8. See Howard, *Garden Cities of To-Morrow*.

9. Howard, *Garden Cities of To-Morrow*, 133–134.

10. Le Corbusier, *The Radiant City: Elements of a Doctrine of Urbanism to Be Used as the Basis of Our Machine-Age Civilization* (New York: Orion Press, 1964), 134, 240, 134.

11. Le Corbusier, *Aircraft: The New Vision* (New York: Studio Publications, 1935), 96, 5, 100.

12. Le Corbusier, *The Radiant City*, 121.

13. Le Corbusier, *The Radiant City*, 27, 29, 116.

14. Scott, *Seeing Like a State*, 348.

15. Le Corbusier, *The Radiant City*, 181, 154.

16. Le Corbusier, *The Radiant City*, 181, 154.

17. James Holston, *The Modernist City: An Anthropological Critique of Brasília* (Chicago: University of Chicago Press, 1989), 168.

18. Holston, *The Modernist City*, 23, 24, 105.

19. Robert A Caro, *The Power Broker: Robert Moses and the Fall of New York* (1974; repr., New York: Random House, 2015), 909.

20. Harrison E. Salisbury, *The Shook-Up Generation* (New York: Harper and Row, 1958), 73, 75.

21. See Peter Marcuse, *Robert Moses and Public Housing: Contradiction In, Contradiction Out* ([New York: P. Marcuse], 1989).

22. James Baldwin, interview by Kenneth Clark, WGBH-TV, May 24, 1963; published in *Conversations with James Baldwin*, ed. Fred L. Standley and Louis H. Pratt (Jackson: University Press of Mississippi, 1989), 42.

23. Jane Jacobs, *The Death and Life of Great American Cities* (1961; repr., New York: Vintage Books, 1992), 4, 428, 222, 447, 222, 439.

24. Jacobs, *The Death and Life of Great American Cities*, 435, 438-439.

25. Jacobs, *The Death and Life of Great American Cities*, 21.

26. Le Corbusier, *The Radiant City*, 202; Senseable City Lab, "DriveWAVE by MIT SENSEable City Lab," http://senseable.mit.edu/wave/.

27. Hitachi, "City of Boston: Smart City RFI Response" (2017), p. 7, https://drive.google.com/file/d/0B_QckxNE_FoEeVJ5amJVT3NEZXc.

28. Living PlanIT, "Living PlanIT—Boston Smart City RFI" (January 2017), p. 1, https://drive.google.com/file/d/0B_QckxNE_FoEVEUtTFB4SDRhc00.

29. Adora Cheung, "New Cities," *Y Combinator Blog* (June 27, 2016), https://blog.ycombinator.com/new-cities/.

30. Cheung, "New Cities."

31. Le Corbusier, *The Radiant City*, 154.

32. Eric Jaffe, "How Are Those Cities of the Future Coming Along?," *CityLab* (September 11, 2013), https://www.citylab.com/life/2013/09/how-are-those-cities -future-coming-along/6855/.

33. Cheung, "New Cities."

34. Dan Doctoroff, quoted in Leslie Hook, "Alphabet Looks for Land to Build Experimental City," *Financial Times*, September 19, 2017, https://www.ft.com/ content/22b45326-9d47-11e7-9a86-4d5a475ba4c5.

35. Sidewalk Labs, "Vision Sections of RFP Submission" (October 17, 2017), p. 15, https://sidewalktoronto.ca/wp-content/uploads/2017/10/Sidewalk-Labs-Vision -Sections-of-RFP-Submission.pdf.

36. Le Corbusier, *The Radiant City*, 181, 154.

37. Doctoroff, "Reimagining Cities from the Internet Up."

38. Bianca Wylie, "Debrief on Sidewalk Toronto Public Meeting #2—Time to Start Over, Extend the Process," Medium (May 6, 2018), https://medium.com/ @biancawylie/sidewalk-toronto-public-meeting-2-time-to-start-over-extend-the -process-a0575b3adfc3.

39. Jascha Franklin-Hodge, in Knight Foundation, "NetGain Internet of Things Conference" (2017), https://www.youtube.com/watch?v=29u1C4Z6PR4.

40. Nigel Jacob, interview by Ben Green, April 7, 2017.

41. Mayor's Office of New Urban Mechanics, "Boston Smart City Playbook" (2016), https://monum.github.io/playbook/.

42. Mayor's Office of New Urban Mechanics, "Boston Smart City Playbook."

43. Franklin-Hodge, in Knight Foundation, "NetGain Internet of Things Conference."

44. Mimi Kirk, "Why Singapore Will Get Self-Driving Cars First," *CityLab* (August 3, 2016), https://www.citylab.com/transportation/2016/08/why-singapore-leads-in-self -driving-cars/494222/; Annabelle Liang and Dee-Ann Durbin, "World's First Self-Driving Taxis Debut in Singapore," *Bloomberg*, August 25, 2016.

45. Abdur Rahman Alfa Shaban, "Ethiopia Bags a Continental First with $2.2m Smart Parking Facility," *Africanews*, June 15, 2017, http://www.africanews.com/2017/ 06/15/ethiopia-s-22m-smart-parking-facility-is-africa-s-first/.

46. Hollie Russon Gilman, *Democracy Reinvented: Participatory Budgeting and Civic Innovation in America* (Washington, DC: Brookings Institution Press, 2016), 7, 36.

47. See Rafael Sampaio and Tiago Peixoto, "Electronic Participatory Budgeting: False Dilemmas and True Complexities," in *Hope for Democracy: 25 Years of Participatory Budgeting Worldwide*, ed. Nelson Dias (São Brás de Alportel, Portugal: In Loco Association, 2014), 413–425.

48. Josh Chin, "About to Break the Law? Chinese Police Are Already On to You," *Wall Street Journal*, February 27, 2018, https://www.wsj.com/articles/china -said-to-deploy-big-data-for-predictive-policing-in-xinjiang-1519719096.

49. Josh Chin and Clément Bürge, "Twelve Days in Xinjiang: How China's Surveillance State Overwhelms Daily Life," *Wall Street Journal*, December 19, 2017, https://www .wsj.com/articles/twelve-days-in-xinjiang-how-chinas-surveillance-state-overwhelms -daily-life-1513700355.

50. Adrian Short, "BT InLink in London: Building a Privatised 'Smart City' by Stealth" (December 14, 2017), https://www.adrianshort.org/posts/2017/bt-inlink -london-smart-city/.

51. Natasha Lomas, "How 'Anonymous' Wifi Data Can Still Be a Privacy Risk," *TechCrunch* (October 7, 2017), https://techcrunch.com/2017/10/07/how -anonymous-wifi-data-can-still-be-a-privacy-risk/.

52. Laura Adler, "How Smart City Barcelona Brought the Internet of Things to Life," *Data-Smart City Solutions* (February 18, 2016), https://datasmart.ash.harvard.edu/ news/article/how-smart-city-barcelona-brought-the-internet-of-things-to-life-789.

53. Albert Canigueral, "In Barcelona, Technology Is a Means to an End for a Smart City," *GreenBiz* (September 12, 2017), https://www.greenbiz.com/article/ barcelona-technology-means-end-smart-city.

54. Carla Bailo, interview by Ben Green, May 9, 2017.

References

Adler, Laura. "How Smart City Barcelona Brought the Internet of Things to Life." *Data-Smart City Solutions* (February 18, 2016). https://datasmart.ash.harvard.edu/news/article/how-smart-city-barcelona-brought-the-internet-of-things-to-life-789.

Aitoro, Jill R. "Defining Privacy Protection by Acknowledging What It's Not." *Federal Times*, March 8, 2016. http://www.federaltimes.com/story/government/interview/one-one/2016/03/08/defining-privacy-protection-acknowledging-what-s-not/81464556/.

Alexander, Michelle. *The New Jim Crow: Mass Incarceration in the Age of Colorblindness*. New York: New Press, 2012.

Altshuler, Alan. *The Urban Transportation System: Politics & Policy Innovations*. Cambridge, MA: MIT Press, 1981.

American Civil Liberties Union. "Community Control over Police Surveillance" (2018). https://www.aclu.org/issues/privacy-technology/surveillance-technologies/community-control-over-police-surveillance.

Anderson, Chris. "The End of Theory: The Data Deluge Makes the Scientific Method Obsolete." *Wired*, June 23, 2008. https://www.wired.com/2008/06/pb-theory/.

Anderson, Elizabeth S. 1999. "What Is the Point of Equality?" *Ethics* 109, no. 2 (1999): 287–337.

Angwin, Julia, Jeff Larson, Surya Mattu, and Lauren Kirchner. "Machine Bias." *ProPublica*, May 23, 2016. https://www.propublica.org/article/machine-bias-risk-assessments-in-criminal-sentencing.

Aquilina, Kimberly M. "50 US Cities Pen Letter to FCC Demanding Net Neutrality, Democracy." *Metro*, July 12, 2017. https://www.metro.us/news/local-news/net-neutrality-50-cities-letter-fcc-democracy.

Array of Things. "Array of Things" (2016). http://arrayofthings.github.io/.

Array of Things. "Array of Things Operating Policies" (August 15, 2016). https://arrayofthings.github.io/final-policies.html.

Array of Things. "Responses to Public Feedback" (2016). https://arrayofthings.github.io/policy-responses.html.

Asher, Jeff, and Rob Arthur. "Inside the Algorithm That Tries to Predict Gun Violence in Chicago." *New York Times*, June 13, 2017. https://www.nytimes.com/2017/06/13/upshot/what-an-algorithm-reveals-about-life-on-chicagos-high-risk-list.html.

Atkinson, Craig, dir. *Do Not Resist*. Vanish Films, 2016.

Azemati, Hanna, and Christina Grover-Roybal. "Shaking Up the Routine: How Seattle Is Implementing Results-Driven Contracting Practices to Improve Outcomes for People Experiencing Homelessness." Harvard Kennedy School Government Performance Lab (September 2016). http://govlab.hks.harvard.edu/files/siblab/files/seattle_rdc_policy_brief_final.pdf.

Backstrom, Lars, Eric Sun, and Cameron Marlow. "Find Me If You Can: Improving Geographical Prediction with Social and Spatial Proximity." Paper presented at the Proceedings of the 19th International Conference on World Wide Web, Raleigh, NC, April 2010.

Badger, Emily. "Pave Over the Subway? Cities Face Tough Bets on Driverless Cars." *New York Times*, July 20, 2018. https://www.nytimes.com/2018/07/20/upshot/driverless-cars-vs-transit-spending-cities.html.

Baker, Al, J. David Goodman and Benjamin Mueller. "Beyond the Chokehold: The Path to Eric Garner's Death." *New York Times*, June 13, 2015. https://www.nytimes.com/2015/06/14/nyregion/eric-garner-police-chokehold-staten-island.html.

Baldwin, James. *Conversations with James Baldwin*. Edited by Fred L. Standley and Louis H. Pratt. Jackson: University Press of Mississippi, 1989.

Bauman, Matthew J., et al. "Reducing Incarceration through Prioritized Interventions." In *COMPASS '18: Proceedings of the 1st ACM SIGCAS Conference on Computing and Sustainable Societies*. 2018.

Baumgardner, Kinder. "Beyond Google's Cute Car: Thinking Through the Impact of Self-Driving Vehicles on Architecture." *Cite: The Architecture + Design Review of Houston* (2015): 36–43.

Bayley, David H. *Police for the Future*. New York: Oxford University Press, 1996.

Beekman, Daniel, and Jack Broom. "Mayor, County Exec Declare 'State of Emergency' over Homelessness." *Seattle Times*, January 31, 2016. http://www.seattletimes.com/seattle-news/politics/mayor-county-exec-declare-state-of-emergency-over-homelessness/.

Bell, Monica C. "Police Reform and the Dismantling of Legal Estrangement." *Yale Law Journal* 126 (2017): 2054–2150.

Berman, Marshall. "Take It to the Streets: Conflict and Community in Public Space." *Dissent* 33, no. 4 (1986): 476–485.

Bertrand, Marianne, and Sendhil Mullainathan. "Are Emily and Greg More Employable Than Lakisha and Jamal? A Field Experiment on Labor Market Discrimination." *American Economic Review* 94, no. 4 (2004): 991–1013.

Bibbins, J. Rowland. "Traffic-Transportation Planning and Metropolitan Development." *ANNALS of the American Academy of Political and Social Science* 116, no. 1 (1924): 205–214.

Bliss, Laura. "Columbus Now Says 'Smart' Rides for Vulnerable Moms Are Coming." *CityLab* (December 1, 2017). https://www.citylab.com/transportation/2017/12/columbus-now-says-smart-rides-for-vulnerable-moms-are-coming/547013/.

Bliss, Laura. "Who Wins When a City Gets Smart?" *CityLab* (November 1, 2017). https://www.citylab.com/transportation/2017/11/when-a-smart-city-doesnt-have-all-the-answers/542976/.

Bonczar, Thomas P. "Prevalence of Imprisonment in the U.S. Population, 1974–2001." *Bureau of Justice Statistics Special Report* (2003). https://www.bjs.gov/content/pub/pdf/piusp01.pdf.

Bond, Robert M., Christopher J. Fariss, Jason J. Jones, Adam D. I. Kramer, Cameron Marlow, Jaime E. Settle, and James H. Fowler. "A 61-Million-Person Experiment in Social Influence and Political Mobilization." *Nature* 489, no. 7415 (2012): 295–298.

Bond-Graham, Darwin, and Ali Winston. "All Tomorrow's Crimes: The Future of Policing Looks a Lot Like Good Branding." *SF Weekly,* October 30, 2013. http://archives.sfweekly.com/sanfrancisco/all-tomorrows-crimes-the-future-of-policing-looks-a-lot-like-good-branding/.

Boston, City of, Mayor's Office of New Urban Mechanics. "Boston Smart City Playbook" (2016). https://monum.github.io/playbook/.

Boston, City of, Mayor's Office of New Urban Mechanics. "Civic Research Agenda" (March 15, 2018). https://www.boston.gov/departments/new-urban-mechanics/civic-research-agenda.

Brauneis, Robert, and Ellen P. Goodman. "Algorithmic Transparency for the Smart City." *Yale Journal of Law and Technology* 20 (2018): 103–176.

Braxton, Karissa. "City's Homeless Response Investments Are Housing More People." *City of Seattle Human Interests Blog* (May 31, 2018). http://humaninterests.seattle.gov/2018/05/31/citys-homeless-response-investments-are-housing-more-people/.

Brazel, Rosalind. "City of Seattle Hires Ginger Armbruster as Chief Privacy Officer." *Tech Talk Blog* (July 11, 2017). http://techtalk.seattle.gov/2017/07/11/city-of-seattle-hires-ginger-armbruster-as-chief-privacy-officer/.

"Brigade." https://www.brigade.com/.

Brustein, Joshua. "This Guy Trains Computers to Find Future Criminals." *Bloomberg* (2016). https://www.bloomberg.com/features/2016-richard-berk-future-crime/.

Buell, Ryan W., Ethan Porter, and Michael I. Norton. "Surfacing the Submerged State: Operational Transparency Increases Trust in and Engagement with Government." Harvard Business School Working Paper No. 14-034 (November 2013; rev. March 2018).

"Buggy Capital of the World." *Columbus Dispatch*, Blog (July 29, 2015). http://www.dispatch.com/content/blogs/a-look-back/2015/07/buggy-capital-of-the-world.html.

Burns, Haskel. "Ordinance Looks at Police Surveillance Equipment." *Hattiesburg American*, October 28, 2016. https://www.hattiesburgamerican.com/story/news/local/hattiesburg/2016/10/28/ordinance-looks-police-surveillance-equipment/92899430/.

Burns, Nancy, Kay Lehman Schlozman, and Sidney Verba. *The Private Roots of Public Action: Gender, Equality, and Political Participation*. Cambridge, MA: Harvard University Press, 2001.

Butler, Paul. *Chokehold: Policing Black Men*. New York: New Press, 2017.

Calthorpe Associates, Mid-Ohio Regional Planning Commission, Columbus District Council of the Urban Land Institute, and Columbus 2020. "insight2050 Scenario Results Report" (February 26, 2015). http://getinsight2050.org/wp-content/uploads/2015/03/2015_02_26-insight2050-Report.pdf.

Canigueral, Albert. "In Barcelona, Technology Is a Means to an End for a Smart City." *GreenBiz* (September 12, 2017). https://www.greenbiz.com/article/barcelona-technology-means-end-smart-city.

Caro, Robert A. *The Power Broker: Robert Moses and the Fall of New York*. 1974. Repr., New York: Random House, 2015.

Center for Data Science and Public Policy, University of Chicago. "Data-Driven Justice Initiative: Identifying Frequent Users of Multiple Public Systems for More Effective Early Assistance" (2018). https://dsapp.uchicago.edu/projects/criminal-justice/data-driven-justice-initiative/.

Chamberlain, Allison T., Jonathan D. Lehnert, and Ruth L. Berkelman. "The 2015 New York City Legionnaires' Disease Outbreak: A Case Study on a History-Making Outbreak." *Journal of Public Health Management and Practice* 23, no. 4 (2017): 410–416.

Chambers, John, and Wim Elfrink. "The Future of Cities." *Foreign Affairs*, October 31, 2014. https://www.foreignaffairs.com/articles/2014-10-31/future-cities.

Chammah, Maurice. "Policing the Future." *The Verge* (2016). http://www.theverge.com/2016/2/3/10895804/st-louis-police-hunchlab-predictive-policing-marshall-project.

Cheung, Adora. "New Cities." *Y Combinator Blog* (June 27, 2016). https://blog.ycombinator.com/new-cities/.

Chicago, City of. "Food Inspection Forecasting" (2017). https://chicago.github.io/food-inspections-evaluation/.

Chiel, Ethan. "Why the D.C. Government Just Publicly Posted Every D.C. Voter's Address Online." *Splinter*, June 14, 2016. https://splinternews.com/why-the-d-c-government-just-publicly-posted-every-d-c-1793857534.

Chin, Josh. "About to Break the Law? Chinese Police Are Already on to You." *Wall Street Journal*, February 27, 2018. https://www.wsj.com/articles/china-said-to-deploy-big-data-for-predictive-policing-in-xinjiang-1519719096.

Chin, Josh, and Clément Bürge. "Twelve Days in Xinjiang: How China's Surveillance State Overwhelms Daily Life." *Wall Street Journal*, December 19, 2017. https://www.wsj.com/articles/twelve-days-in-xinjiang-how-chinas-surveillance-state-overwhelms-daily-life-1513700355.

Civic Analytics Network. "An Open Letter to the Open Data Community." *Data-Smart City Solutions*, March 3, 2017. https://datasmart.ash.harvard.edu/news/article/an-open-letter-to-the-open-data-community-988.

Claypool, Henry, Amitai Bin-Nun, and Jeffrey Gerlach. "Self-Driving Cars: The Impact on People with Disabilities." *The Ruderman White Paper* (January 2017). http://secureenergy.org/wp-content/uploads/2017/01/Self-Driving-Cars-The-Impact-on-People-with-Disabilities_FINAL.pdf.

Columbus, City of. "Columbus Smart City Application" (2016). https://www.transportation.gov/sites/dot.gov/files/docs/Columbus OH Vision Narrative.pdf.

Columbus, City of. "Linden Infant Mortality Profile" (2018). http://celebrateone.info/wp-content/uploads/2018/03/Linden_IMProfile_9.7.pdf.

Cushing, Tim. "'Predictive Policing' Company Uses Bad Stats, Contractually-Obligated Shills to Tout Unproven 'Successes.'" *Techdirt*, November 1, 2013. https://www.techdirt.com/articles/20131031/13033125091/predictive-policing-company-uses-bad-stats-contractually-obligated-shills-to-tout-unproven-successes.shtml.

Cutler, Kim-Mai, and Josh Constine. "Sean Parker's Brigade App Enters Private Beta as a Dead-Simple Way of Taking Political Positions." *TechCrunch*, June 17, 2015. https://techcrunch.com/2015/06/17/sean-parker-brigade/.

Daly, Jimmy. "10 Cities with 311 iPhone Applications." *StateTech*, August 10, 2012. https://statetechmagazine.com/article/2012/08/10-cities-311-iphone-applications.

Dastin, Jeffrey. "Amazon Scraps Secret AI Recruiting Tool that Showed Bias Against Women." *Reuters*, October 9, 2018. https://www.reuters.com/article/us-amazon-com-jobs-automation-insight/amazon-scraps-secret-ai-recruiting-tool-that-showed-bias-against-women-idUSKCN1MK08G.

Data-Driven Justice Initiative. "Data-Driven Justice Playbook" (2016). http://www.naco.org/sites/default/files/documents/DDJ Playbook Discussion Draft 12.8.16_1.pdf.

"Data-Driven Research, Policy, and Practice: Friday Opening Remarks." Boston Area Research Initiative, March 10, 2017. https://www.youtube.com/watch?v=cRINlFFBHBo.

DataSF. "Data Quality" (2017). https://datasf.org/resources/data-quality/.

DataSF. "Data Standards Reference Handbook" (2018). https://datasf.gitbooks.io/draft-publishing-standards/.

DataSF. "DataScienceSF" (2017). https://datasf.org/science/.

DataSF. "DataSF in Progress" (2018). https://datasf.org/progress/.

DataSF. "Eviction Alert System" (2018). https://datasf.org/showcase/datascience/eviction-alert-system/.

DataSF. "Keeping Moms and Babies in Nutrition Program" (2018). https://datasf.org/showcase/datascience/keeping-moms-and-babies-in-nutrition-program/.

de Montjoye, Yves-Alexandre, César A. Hidalgo, Michel Verleysen, and Vincent D. Blondel. "Unique in the Crowd: The Privacy Bounds of Human Mobility." *Scientific Reports* 3, art. no. 1376 (2013). https://doi.org/10.1038/srep01376.

de Montjoye, Yves-Alexandre, Laura Radaelli, Vivek Kumar Singh, and Alex "Sandy" Pentland. "Unique in the Shopping Mall: On the Reidentifiability of Credit Card Metadata." *Science* 347, no. 6221 (2015): 536–539.

De Tocqueville, Alexis. *Democracy in America*. Edited by Max Lerner and J.-P. Mayer, trans. George Lawrence. 2 vols. New York: Harper and Row, 1966.

Desmond, Matthew, Andrew V. Papachristos, and David S. Kirk. "Police Violence and Citizen Crime Reporting in the Black Community." *American Sociological Review* 81, no. 5 (2016): 857–876.

Detroit Future City. "2012 Detroit Strategic Framework Plan" (2012). https://detroitfuturecity.com/wp-content/uploads/2014/12/DFC_Full_2nd.pdf.

Dewey, John. *Logic: The Theory of Inquiry*. New York: H. Holt and Company, 1938.

Diehl, Phil. "Malware Blamed for City's Data Breach." *San Diego Tribune*, September 12, 2017. http://www.sandiegouniontribune.com/communities/north-county/sd-no-malware-letter-20170912-story.html.

Doctoroff, Daniel L. "Reimagining Cities from the Internet Up." Medium: Sidewalk Talk (November 30, 2016). https://medium.com/sidewalk-talk/reimagining-cities-from-the-internet-up-5923d6be63ba.

Downs, Anthony. "The Law of Peak-Hour Expressway Congestion." *Traffic Quarterly* 16, no. 3 (1962): 393–409.

Downs, Anthony. *Still Stuck in Traffic: Coping with Peak-Hour Traffic Congestion.* Washington, DC: Brookings Institution Press, 2005.

Downs, Anthony. "Traffic: Why It's Getting Worse, What Government Can Do." Brookings Institution Policy Brief #128 (January 2004).

Duranton, Gilles, and Matthew A. Turner. "The Fundamental Law of Road Congestion: Evidence from US Cities." *American Economic Review* 101, no. 6 (2011): 2616–2652.

Eagle, Nathan, Alex Sandy Pentland, and David Lazer. "Inferring Friendship Network Structure by Using Mobile Phone Data." *Proceedings of the National Academy of Sciences* 106, no. 36 (2009): 15274–15278.

Efrati, Amir. "Uber Finds Deadly Accident Likely Caused by Software Set to Ignore Objects on Road." *The Information*, May 7, 2018. https://www.theinformation.com/articles/uber-finds-deadly-accident-likely-caused-by-software-set-to-ignore-objects-on-road.

Engelhardt, Will, Risë Haneberg, Rob MacDougall, Chris Schneweis, and Imran Chaudhry. "Sharing Information between Behavioral Health and Criminal Justice Systems." Council of State Governments Justice Center (March 31, 2016). https://csgjusticecenter.org/wp-content/uploads/2016/03/JMHCP-Info-Sharing-Webinar.pdf.

Eubanks, Virginia. *Automating Inequality: How High-Tech Tools Profile, Police, and Punish the Poor.* New York: St. Martin's Press, 2018.

Eubanks, Virginia. "Technologies of Citizenship: Surveillance and Political Learning in the Welfare System." In *Surveillance and Security: Technological Politics and Power in Everyday Life*, edited by Torin Monahan. New York: Routledge, 2006.

Fagnant, Daniel J., and Kara Kockelman. "Preparing a Nation for Autonomous Vehicles: Opportunities, Barriers and Policy Recommendations." *Transportation Research Part A: Policy and Practice* 77 (2015): 167–181.

Falconer, Gordon, and Shane Mitchell. "Smart City Framework: A Systematic Process for Enabling Smart+Connected Communities" (2012). https://www.cisco.com/c/dam/en_us/about/ac79/docs/ps/motm/Smart-City-Framework.pdf.

Fayyad, Abdallah. "The Criminalization of Gentrifying Neighborhoods." *The Atlantic*, December 20, 2017. https://www.theatlantic.com/politics/archive/2017/12/the-criminalization-of-gentrifying-neighborhoods/548837/.

Federal Trade Commission. *Data Brokers: A Call for Transparency and Accountability.* Washington, DC: Federal Trade Commission, 2014.

Feigenbaum, James J., and Andrew Hall. "How High-Income Areas Receive More Service from Municipal Government: Evidence from City Administrative Data" (2015). https://ssrn.com/abstract=2631106.

Feldman, Andrew, and Jason Johnson, "How Better Procurement Can Drive Better Outcomes for Cities." *Governing*, October 12, 2017. http://www.governing.com/gov -institute/voices/col-cities-3-steps-procurement-reform-better-outcomes.html.

Ferenstein, Greg. "Brigade: New Social Network from Facebook Co-founder Aims to 'Repair Democracy.'" *The Guardian*, June 17, 2015. https://www.theguardian.com/ media/2015/jun/17/brigade-social-network-voter-turnout-sean-parker.

Ferenstein Wire. "Sean Parker Explains His Plans to 'Repair Democracy' with a New Social Network." *Fast Company* (2015). https://www.fastcompany.com/3047571/ sean-parker-explains-his-plans-to-repair-democracy-with-a-new-social-network.

Ferguson, Andrew G. "The Allure of Big Data Policing." *PrawfsBlawg* (May 25, 2017). http://prawfsblawg.blogs.com/prawfsblawg/2017/05/the-allure-of-big-data-policing. html.

Forrest, Adam. "Detroit Battles Blight through Crowdsourced Mapping Project." *Forbes*, June 22, 2015. https://www.forbes.com/sites/adamforrest/2015/06/22/detroit -battles-blight-through-crowdsourced-mapping-project.

Fountain, Jane E. "Paradoxes of Public Sector Customer Service." *Governance* 14, no. 1 (2001): 55–73.

Friess, Steve. "Son of Dem Royalty Creates a Ruck.us." *Politico*, June 26, 2012. http://www.politico.com/story/2012/06/son-of-democratic-party-royalty-creates -a-ruckus-077847.

Fry, Christopher, and Henry Lieberman. *Why Can't We All Just Get Along?* Self-published, 2018. https://www.whycantwe.org/.

Fung, Archon, Hollie Russon Gilman, and Jennifer Shkabatur. "Six Models for the Internet + Politics." *International Studies Review* 15, no. 1 (2013): 30–47.

Furth, Peter. "Pedestrian-Friendly Traffic Signal Timing Policy Recommendations." *Boston City Council Committee on Parks, Recreation, and Transportation* (December 6, 2016). http://www.northeastern.edu/peter.furth/wp-content/uploads/2016/12/ Pedestrian-Friendly-Traffic-Signal-Policies-Boston.pdf.

Garlick, Ross. "Privacy Inequality Is Coming, and It Does Not Look Pretty." *Fordham Political Review*, March 17, 2015. http://fordhampoliticalreview.org/privacy-inequality -is-coming-and-it-does-not-look-pretty/.

General Motors. "To New Horizons" (1939). Posted as "Futurama at 1939 NY World's Fair." https://www.youtube.com/watch?v=sClZqfnWqmc.

Gilliom, John. *Overseers of the Poor: Surveillance, Resistance, and the Limits of Privacy.* Chicago: University of Chicago Press, 2001.

Gilman, Hollie Russon. *Democracy Reinvented: Participatory Budgeting and Civic Innovation in America.* Washington, DC: Brookings Institution Press, 2016.

Glaser, April. "Sanctuary Cities Are Handing ICE a Map." *Slate,* March 13, 2018. https:// slate.com/technology/2018/03/how-ice-may-be-able-to-access-license-plate-data -from-sanctuary-cities-and-use-it-for-arrests.html.

Goel, Sharad, Justin M. Rao, and Ravi Shroff. "Precinct or Prejudice? Understanding Racial Disparities in New York City's Stop-and-Frisk Policy." *Annals of Applied Statistics* 10, no. 1 (2016): 365–394.

Gordon, Eric. "Civic Technology and the Pursuit of Happiness." *Governing* (2016). http://www.governing.com/cityaccelerator/civic-technology-and-the-pursuit-of -happiness.html.

Gordon, Eric, and Jessica Baldwin-Philippi. "Playful Civic Learning: Enabling Lateral Trust and Reflection in Game-Based Public Participation." *International Journal of Communication* 8 (2014): 759–786.

Gordon, Eric, and Stephen Walter. "Meaningful Inefficiencies: Resisting the Logic of Technological Efficiency in the Design of Civic Systems." In *Civic Media: Technology, Design, Practice,* edited by Eric Gordon and Paul Mihailidis. Cambridge, MA: MIT Press, 2016.

Gorner, Jeremy. "With Violence Up, Chicago Police Focus on a List of Likeliest to Kill, Be Killed." *Chicago Tribune,* July 22, 2016. http://www.chicagotribune.com/news/ ct-chicago-police-violence-strategy-met-20160722-story.html.

Graham, Thomas. "Barcelona Is Leading the Fightback against Smart City Surveillance." *Wired,* May 18, 2018. http://www.wired.co.uk/article/barcelona-decidim-ada -colau-francesca-bria-decode.

Green, Ben, Gabe Cunningham, Ariel Ekblaw, Paul Kominers, Andrew Linzer, and Susan Crawford. "Open Data Privacy: A Risk-Benefit, Process-Oriented Approach to Sharing and Protecting Municipal Data." *Berkman Klein Center for Internet & Society Research Publication* (2017). http://nrs.harvard.edu/urn-3:HUL.InstRepos:30340010.

Green, Ben, Thibaut Horel, and Andrew V. Papachristos. 2017. "Modeling Contagion through Social Networks to Explain and Predict Gunshot Violence in Chicago, 2006 to 2014." *JAMA Internal Medicine* 177, no. 3 (2017): 326–333. https://doi.org/10 .1001/jamainternmed.2016.8245.

Greenfield, Adam. *Against the Smart City*. New York: Do Projects, 2013.

Griffin, Dominic. "'Do Not Resist' Traces the Militarization of Police with Unprecedented Access to Raids and Unrest." *Baltimore City Paper*, November 1, 2016. http://www.citypaper.com/film/film/bcp-110216-screens-do-not-resist-20161101 -story.html.

Guynn, Jessica. "ACLU: Police Used Twitter, Facebook to Track Protests." *USA Today*, October 12, 2016. https://www.usatoday.com/story/tech/news/2016/10/11/aclu -police-used-twitter-facebook-data-track-protesters-baltimore-ferguson/91897034/.

Ha, Anthony. "Sean Parker: Defeating SOPA Was the 'Nerd Spring.'" *TechCrunch*, March 12, 2012. https://techcrunch.com/2012/03/12/sean-parker-defeating-sopa-was -the-nerd-spring/.

Halper, Evan. "Napster Co-founder Sean Parker Once Vowed to Shake Up Washington—So How's That Working Out?" *Los Angeles Times*, August 4, 2016. http://www .latimes.com/politics/la-na-pol-sean-parker-20160804-snap-story.html.

Han, Hahrie. *How Organizations Develop Activists: Civic Associations and Leadership in the 21st Century*. New York: Oxford University Press, 2014.

Harcourt, Bernard E. "Risk as a Proxy for Race: The Dangers of Risk Assessment." *Federal Sentencing Reporter* 27, no. 4 (2015): 237–243.

Hartnett, Kevin. "Bye-bye Traffic Lights." *Boston Globe*, March 28, 2016. https:// www.bostonglobe.com/ideas/2016/03/28/bye-bye-traffic-lights/8HSV9DZa4qPC1tH 4zQ4pTO/story.html.

Henry, Jason. "Los Angeles Police, Sheriff's Scan over 3 Million License Plates a Week." *San Gabriel Valley Tribune*, August 26, 2014. https://www.sgvtribune.com/2014/08/26/ los-angeles-police-sheriffs-scan-over-3-million-license-plates-a-week/.

Herrold, George. "City Planning and Zoning." *Canadian Engineer* 45 (1923): 128–130.

Herrold, George. "The Parking Problem in St. Paul." *Nation's Traffic* 1 (July 1927): 28–30, 47–48.

Herz, Ansel. "How the Seattle Police Secretly—and Illegally—Purchased a Tool for Tracking Your Social Media Posts." *The Stranger*, September 28, 2016. https://www .thestranger.com/news/2016/09/28/24585899/how-the-seattle-police-secretlyand -illegallypurchased-a-tool-for-tracking-your-social-media-posts.

Hitachi. "City of Boston: Smart City RFI Response" (2017). https://drive.google.com/ file/d/0B_QckxNE_FoEeVJ5amJVT3NEZXc.

Holston, James. *The Modernist City: An Anthropological Critique of Brasília*. Chicago: University of Chicago Press, 1989.

Hook, Leslie. "Alphabet Looks for Land to Build Experimental City." *Financial Times*, September 19, 2017. https://www.ft.com/content/22b45326-9d47-11e7 -9a86-4d5a475ba4c5.

Howard, Ebenezer. *Garden Cities of To-Morrow*. London: Swan Sonnenschein, 1902.

Hughes, John, dir. *Ferris Bueller's Day Off*. Paramount Pictures, 1986.

HunchLab. "Next Generation Predictive Policing." https://www.hunchlab.com/.

Hunt, Priscillia, Jessica Saunders, and John S. Hollywood. *Evaluation of the Shreveport Predictive Policing Experiment*. RR-531-NIJ. Santa Monica, CA: RAND Corporation, 2014.

Hvistendahl, Mara. "Can 'Predictive Policing' Prevent Crime Before It Happens?" *Science*, September 28, 2016. http://www.sciencemag.org/news/2016/09/can -predictive-policing-prevent-crime-it-happens.

IBM. "Predictive Analytics: Police Use Analytics to Reduce Crime" (2012). https:// www.youtube.com/watch?v=iY3WRvXVogo.

IBM. "What Is a Self-Service Government?" *The Atlantic* [advertising]. http://www.the atlantic.com/sponsored/ibm-transformation/what-is-a-self-service-government/248/.

Jacobs, Jane. *The Death and Life of Great American Cities*. 1961. Repr., New York: Vintage Books, 1992.

Jaffe, Eric. "How Are Those Cities of the Future Coming Along?" *CityLab* (September 11, 2013). https://www.citylab.com/life/2013/09/how-are-those-cities-future-coming -along/6855/.

James, Letitia, and Ben Kallos. "New York City Digital Divide Fact Sheet." Press release, March 16, 2017.

Jenkins, J. L. "Illegal Parking Hinders Work of Stop-Go Lights; Pedestrian Dangers Grow as Loop Speeds Up." *Chicago Tribune*, February 10, 1926.

Joh, Elizabeth E. "The Undue Influence of Surveillance Technology Companies on Policing." *New York University Law Review* 92 (2017): 101–130.

Johnston, Casey. "Denied for That Loan? Soon You May Thank Online Data Collection." *ArsTechnica* (2013). https://arstechnica.com/business/2013/10/denied-for -that-loan-soon-you-may-thank-online-data-collection/.

Joseph, George. "Exclusive: Feds Regularly Monitored Black Lives Matter Since Ferguson." *The Intercept*, July 24, 2015. https://theintercept.com/2015/07/24/ documents-show-department-homeland-security-monitoring-black-lives-matter -since-ferguson/.

Jouvenal, Justin. "The New Way Police Are Surveilling You: Calculating Your Threat 'Score.'" *Washington Post*, January 10, 2016. https://www.washingtonpost.com/local/public-safety/the-new-way-police-are-surveilling-you-calculating-your-threat-score/2016/01/10/e42bccac-8e15-11e5-baf4-bdf37355da0c_story.html.

Jouvenal, Justin. "Police Are Using Software to Predict Crime. Is It a 'Holy Grail' or Biased against Minorities?" *Washington Post*, November 17, 2016. https://www.washingtonpost.com/local/public-safety/police-are-using-software-to-predict-crime-is-it-a-holy-grail-or-biased-against-minorities/2016/11/17/525a6649-0472-440a-aae1-b283aa8e5de8_story.html.

Kang, Cecilia. "Unemployed Detroit Residents Are Trapped by a Digital Divide." *New York Times*, May 23, 2016. https://www.nytimes.com/2016/05/23/technology/unemployed-detroit-residents-are-trapped-by-a-digital-divide.html.

Kang, Cecilia. "Where Self-Driving Cars Go to Learn." *New York Times*, November 11, 2017. https://www.nytimes.com/2017/11/11/technology/arizona-tech-industry-favorite-self-driving-hub.html.

Kaste, Martin. "Orlando Police Testing Amazon's Real-Time Facial Recognition." *National Public Radio*, May 22, 2018. https://www.npr.org/2018/05/22/613115969/orlando-police-testing-amazons-real-time-facial-recognition.

Kayyali, Dia. "The History of Surveillance and the Black Community." *Electronic Frontier Foundation* (February 13, 2014). https://www.eff.org/deeplinks/2014/02/history-surveillance-and-black-community.

Kiley, Brendan, and Matt Fikse-Verkerk. "You Are a Rogue Device." *The Stranger*, November 6, 2013. http://www.thestranger.com/seattle/you-are-a-rogue-device/Content?oid=18143845.

Kirk, Mimi. "Why Singapore Will Get Self-Driving Cars First." *CityLab* (August 3, 2016). https://www.citylab.com/transportation/2016/08/why-singapore-leads-in-self-driving-cars/494222/.

Klarreich, Erica. "Privacy by the Numbers: A New Approach to Safeguarding Data." *Quanta Magazine*, December 10, 2012. https://www.quantamagazine.org/a-mathematical-approach-to-safeguarding-private-data-20121210/.

Kleinberg, Jon, Sendhil Mullainathan, and Manish Raghavan. "Inherent Trade-Offs in the Fair Determination of Risk Scores." *arXiv.org* (2016). https://arxiv.org/abs/1609.05807.

Klockars, Carl B. "Some Really Cheap Ways of Measuring What Really Matters." In *Measuring What Matters: Proceedings from the Policing Research Institute Meetings*. Washington, DC: National Institute of Justice, 1999.

Knight Foundation. "NetGain Internet of Things Conference" (2017). https://www.youtube.com/watch?v=29u1C4Z6PR4.

Kofman, Ava. "Real-Time Face Recognition Threatens to Turn Cops' Body Cameras into Surveillance Machines." *The Intercept*, March 22, 2017. https://theintercept.com/2017/03/22/real-time-face-recognition-threatens-to-turn-cops-body-cameras-into-surveillance-machines/.

Kosinski, Michal, David Stillwell, and Thore Graepel. "Private Traits and Attributes Are Predictable from Digital Records of Human Behavior." *Proceedings of the National Academy of Sciences* 110, no. 15 (2013): 5802–5805.

Kramer, Adam D. I., Jamie E. Guillory, and Jeffrey T. Hancock. "Experimental Evidence of Massive-Scale Emotional Contagion through Social Networks." *Proceedings of the National Academy of Sciences* 111, no. 24 (June 17, 2014): 8788–8790.

Lantos-Swett, Chante. "Leveraging Technology to Improve Participation: Textizen and Oregon's Kitchen Table." *Challenges to Democracy*, Blog (April 4, 2016). http://www.challengestodemocracy.us/home/leveraging-technology-to-improve-participation-textizen-and-oregons-kitchen-table/.

Larson, Selena. "Uber's Massive Hack: What We Know." *CNN*, November 22, 2017. http://money.cnn.com/2017/11/22/technology/uber-hack-consequences-cover-up/index.html.

Latour, Bruno. *The Pasteurization of France*. Translated by Alan Sheridan and John Law. Cambridge, MA: Harvard University Press, 1993.

Latour, Bruno. "Tarde's Idea of Quantification." In *The Social After Gabriel Tarde: Debates and Assessments*, edited by Mattei Candea. London: Routledge, 2010.

Lauper, Cyndi. "Girls Just Want To Have Fun (Official Video)" (1983). https://www.youtube.com/watch?v=PIb6AZdTr-A.

Laura and John Arnold Foundation. "Laura and John Arnold Foundation to Continue Data-Driven Criminal Justice Effort Launched under the Obama Administration." Press release, January 23, 2017. http://www.arnoldfoundation.org/laura-john-arnold-foundation-continue-data-driven-criminal-justice-effort-launched-obama-administration/.

Lavigne, Sam, Brian Clifton, and Francis Tseng. "Predicting Financial Crime: Augmenting the Predictive Policing Arsenal." *The New Inquiry* (2017). https://whitecollar.thenewinquiry.com/static/whitepaper.pdf.

Le Corbusier. *Aircraft: The New Vision*. New York: Studio Publications, 1935.

Le Corbusier. *The Radiant City: Elements of a Doctrine of Urbanism to Be Used as the Basis of Our Machine-Age Civilization*. New York: Orion Press, 1964.

Leadership Conference on Civil and Human Rights, and Upturn. "Police Body Worn Cameras: A Policy Scorecard" (November 2017). https://www.bwcscorecard.org/.

Lerman, Jonas. "Big Data and Its Exclusions." *Stanford Law Review* 66 (2013): 55–63.

Liang, Annabelle, and Dee-Ann Durbin. "World's First Self-Driving Taxis Debut in Singapore." *Bloomberg*, August 25, 2016. https://www.bloomberg.com/news/articles/2016-08-25/world-s-first-self-driving-taxis-debut-in-singapore.

Liebman, Jeff. "Transforming the Culture of Procurement in State and Local Government." Interview by Andy Feldmanm, *Gov Innovator* podcast (April 20, 2017). http://govinnovator.com/jeffrey_liebman_2017/.

LinkNYC. "Find a Link." https://www.link.nyc/find-a-link.html.

LinkNYC. "Privacy Policy" (March 17, 2017). https://www.link.nyc/privacy-policy.html.

Linn, Denise, and Glynis Startz. "Array of Things Civic Engagement Report" (August 2016). https://arrayofthings.github.io/engagement-report.html.

Lipsky, Michael. *Street-Level Bureaucracy: Dilemmas of the Individual in Public Services.* 30th anniversary ed. New York: Russell Sage Foundation, 2010.

Living PlanIT. "Living PlanIT—Boston Smart City RFI" (January 2017). https://drive.google.com/file/d/0B_QckxNE_FoEVEUtTFB4SDRhc00.

Lomas, Natasha. "How 'Anonymous' Wifi Data Can Still Be a Privacy Risk." *TechCrunch*, October 7, 2017. https://techcrunch.com/2017/10/07/how-anonymous-wifi-data-can-still-be-a-privacy-risk/.

Lowry, Stella, and Gordon Macpherson. "A Blot on the Profession." *British Medical Journal* 296, no. 6623 (1988): 657–658.

Lubell, Sam. "Here's How Self-Driving Cars Will Transform Your City." *Wired*, October 21, 2016. https://www.wired.com/2016/10/heres-self-driving-cars-will-transform-city/.

Lum, Kristian. "Predictive Policing Reinforces Police Bias." Human Rights Data Analysis Group (2016). https://hrdag.org/2016/10/10/predictive-policing-reinforces-police-bias/.

Lum, Kristian, and William Isaac. "To Predict and Serve?" *Significance* 13, no. 5 (2016): 14–19.

Madden, Mary, Michele E. Gilman, Karen Levy, and Alice E. Marwick. "Privacy, Poverty and Big Data: A Matrix of Vulnerabilities for Poor Americans." *Washington University Law Review* 95 (2017): 53–125.

Marcuse, Peter. *Robert Moses and Public Housing: Contradiction In, Contradiction Out.* [New York: P. Marcuse], 1989.

Mashariki, Amen Ra. "NYC Data Analytics." Presentation at Esri Senior Executive Summit, 2017. https://www.youtube.com/watch?v=ws8EQg5YlrY.

"Massachusetts Ave & Columbus Ave." Walk Score (2018). https://www.walkscore.com/score/columbus-ave-and-massachusetts-ave-boston.

McFarland, Matt. "Chicago Gets Serious about Tracking Air Quality and Traffic Data." *CNN*, August 29, 2016. http://money.cnn.com/2016/08/29/technology/chicago-sensors-data/index.html.

Menino, Thomas M. "Inaugural Address" (January 4, 2010). https://www.cityofboston.gov/TridionImages/2010%20Thomas%20M%20%20Menino%20Inaugural%20final_tcm1-4838.pdf.

Metcalf, Jacob. "Ethics Review for Pernicious Feedback Loops." Medium: Data & Society: Points (November 7, 2016). https://points.datasociety.net/ethics-review-for-pernicious-feedback-loops-9a7ede4b610e.

Metz, David. *The Limits to Travel: How Far Will You Go?* New York: Routledge, 2012.

"Middlesex Police Discuss Data-Driven Justice Initiative." *Wicked Local Arlington*, December 30, 2016. http://arlington.wickedlocal.com/news/20161230/middlesex-police-discuss-data-driven-justice-initiative.

Misa, Thomas J. "Controversy and Closure in Technological Change. Constructing 'Steel.'" In *Shaping Technology / Building Society: Studies in Sociotechnical Change*, edited by Wiebe E. Bijker and John Law. Cambridge, MA: MIT Press, 1992.

Morozov, Evgeny. *To Save Everything, Click Here: The Folly of Technological Solutionism*. New York: PublicAffairs, 2014.

Moskos, Peter. *Cop in the Hood: My Year Policing Baltimore's Eastern District*. Princeton, NJ: Princeton University Press, 2008.

Murgado, Amaury. "Developing a Warrior Mindset." *POLICE Magazine*, May 24, 2012. http://www.policemag.com/channel/patrol/articles/2012/05/warrior-mindset.aspx.

National Association of Counties. "Data-Driven Justice: Disrupting the Cycle of Incarceration" [2015]. http://www.naco.org/resources/signature-projects/data-driven-justice.

National Association of Counties. "Mental Health and Criminal Justice Case Study: Johnson County, Kan." (2015). http://www.naco.org/sites/default/files/documents/Johnson County Mental Health and Jails Case Study_FINAL.pdf.

National Center for Statistics and Analysis, National Highway Traffic Safety Administration. "Critical Reasons for Crashes Investigated in the National Motor Vehicle Crash Causation Survey." Traffic Safety Facts: Crash Stats, Report No. DOT HS 812 115 (February 2015).

National Center for Statistics and Analysis, National Highway Traffic Safety Administration. "2015 Motor Vehicle Crashes: Overview." Traffic Safety Facts Research Note, Report No. DOT HS 812 318 (August 2016).

National League of Cities. "Trends in Smart City Development" (2016). http://www.nlc.org/sites/default/files/2017-01/Trends in Smart City Development.pdf.

Needham, Catherine. *Citizen-Consumers: New Labour's Marketplace Democracy.* London: Catalyst, 2003.

Needham, Catherine E. "Customer Care and the Public Service Ethos." *Public Administration* 84, no. 4 (2006): 845–860.

New York City Council. "Int 1696-2017: Automated Decision Systems Used by Agencies." (2017). http://legistar.council.nyc.gov/LegislationDetail.aspx?ID=3137815 &GUID=437A6A6D-62E1-47E2-9C42-461253F9C6D0.

New York City Council. "Transcript of the Minutes of the Committee on Technology" (October 16, 2017). http://legistar.council.nyc.gov/View.ashx?M=F&ID=5522569 &GUID=DFECA4F2-E157-42AB-B598-BA3A8185E3FF.

New York City Mayor's Office of Operations. "Hurricane Sandy Response." *NYC Customer Service Newsletter* 5, no. 2 (February 2013). https://www1.nyc.gov/assets/operations/downloads/pdf/nyc_customer_service_newsletter_volume_5_issue_2.pdf.

New York City Office of the Mayor. "Mayor de Blasio Announces Public Launch of LinkNYC Program, Largest and Fastest Free Municipal Wi-Fi Network in the World" (February 18, 2016). http://www1.nyc.gov/office-of-the-mayor/news/184-16/mayor -de-blasio-public-launch-linknyc-program-largest-fastest-free-municipal#/0.

New York Civil Liberties Union. "City's Public Wi-Fi Raises Privacy Concerns" (2016). https://www.nyclu.org/en/press-releases/nyclu-citys-public-wi-fi-raises-privacy -concerns.

New York Times Editorial Board. "The Racism at the Heart of Flint's Crisis." *New York Times*, March 25, 2016. https://www.nytimes.com/2016/03/25/opinion/the-racism -at-the-heart-of-flints-crisis.html.

Newsom, Gavin, and Lisa Dickey. *Citizenville: How to Take the Town Square Digital and Reinvent Government.* New York: Penguin, 2014.

"Nine Additional Cities Join Johnson County's Co-responder Program." Press release, Johnson County, Kansas, July 18, 2016. https://jocogov.org/press-release/nine -additional-cities-join-johnson-county's-co-responder-program.

Norton, Peter D. *Fighting Traffic: The Dawn of the Motor Age in the American City.* Cambridge, MA: MIT Press, 2011.

O'Brien, Daniel Tumminelli, Eric Gordon, and Jessica Baldwin. "Caring about the Community, Counteracting Disorder: 311 Reports of Public Issues as Expressions of Territoriality." *Journal of Environmental Psychology* 40 (2014): 320–330.

O'Brien, Daniel Tumminelli, Dietmar Offenhuber, Jessica Baldwin-Philippi, Melissa Sands, and Eric Gordon. "Uncharted Territoriality in Coproduction: The Motivations for 311 Reporting." *Journal of Public Administration Research and Theory* 27, no. 2 (2017): 320–335.

O'Malley, Nick. "To Predict and to Serve: The Future of Law Enforcement." *Sydney Morning Herald*, March 30, 2013. http://www.smh.com.au/world/to-predict-and-to -serve-the-future-of-law-enforcement-20130330-2h0rb.html.

O'Neil, Cathy. "Don't Grade Teachers with a Bad Algorithm." *Bloomberg*, May 15, 2017. https://www.bloomberg.com/view/articles/2017-05-15/don-t-grade-teachers-with -a-bad-algorithm.

Offenhuber, Dietmar. "The Designer as Regulator: Design Patterns and Categorization in Citizen Feedback Systems." Paper delivered at the Workshop on Big Data and Urban Informatics, Chicago, 2014.

Overmann, Lynn. "Launching the Data-Driven Justice Initiative: Disrupting the Cycle of Incarceration." *The Obama White House* (2016). https://medium.com/ @ObamaWhiteHouse/launching-the-data-driven-justice-initiative-disrupting-the -cycle-of-incarceration-e222448a64cf.

Ozer, Nicole. "Police Use of Social Media Surveillance Software Is Escalating, and Activists Are in the Digital Crosshairs." Medium: ACLU of Northern CA (2016). https://medium.com/@ACLU_NorCal/police-use-of-social-media-surveillance -software-is-escalating-and-activists-are-in-the-digital-d29d8f89c48.

Palmisano, Samuel J. "Smarter Cities: Crucibles of Global Progress." Address, Rio de Janeiro, November 9, 2011. https://www.ibm.com/smarterplanet/us/en/smarter_ cities/article/rio_keynote.html.

Papachristos, Andrew V. "CPD's Crucial Choice: Treat Its List as Offenders or as Potential Victims?" *Chicago Tribune*, July 29, 2016. http://www.chicagotribune.com/ news/opinion/commentary/ct-gun-violence-list-chicago-police-murder-perspec -0801-jm-20160729-story.html.

Papachristos, Andrew V., and Christopher Wildeman. "Network Exposure and Homicide Victimization in an African American Community." *American Journal of Public Health* 104, no. 1 (2014): 143–150.

Pasquale, Frank. "Secret Algorithms Threaten the Rule of Law." *MIT Technology Review*, June 1, 2017. https://www.technologyreview.com/s/608011/secret-algorithms -threaten-the-rule-of -law/.

Perez, Thomas E. "Investigation of the Miami-Dade County Jail." U.S. Department of Justice, Civil Rights Division (August 24, 2011). https://www.clearinghouse.net/chDocs/public/JC-FL-0021-0004.pdf.

Peterson, Andrea. "Why the Names of Six People Who Complained of Sexual Assault Were Published Online by Dallas Police." *Washington Post*, April 29, 2016. https://www.washingtonpost.com/news/the-switch/wp/2016/04/29/why-the-names-of-six-people-who-complained-of-sexual-assault-were-published-online-by-dallas-police/.

Philadelphia, City of. "Open Budget." http://www.phila.gov/openbudget/.

Pinch, Trevor J, and Wiebe E. Bijker. "The Social Construction of Facts and Artifacts: Or How the Sociology of Science and the Sociology of Technology Might Benefit Each Other." In *The Social Construction of Technological Systems: New Directions in the Sociology and History of Technology*, edited by Wiebe E. Bijker, Thomas P. Hughes, and Trevor Pinch. Cambridge, MA: MIT Press, 1987.

Pinto, Nick. "Google Is Transforming NYC's Payphones into a 'Personalized Propaganda Engine.'" *Village Voice*, July 6, 2016. https://www.villagevoice.com/2016/07/06/google-is-transforming-nycs-payphones-into-a-personalized-propaganda-engine/.

Podesta, John, Penny Pritzker, Ernest J. Moniz, John Holdren, and Jeffrey Zients. *Big Data: Seizing Opportunities, Preserving Values*. Washington, DC: Executive Office of the President, 2014.

Porter, Theodore M. *Trust in Numbers: The Pursuit of Objectivity in Science and Public Life*. Princeton, NJ: Princeton University Press, 1995.

Powles, Julia. "New York City's Bold, Flawed Attempt to Make Algorithms Accountable." *New Yorker*, December 20, 2017. https://www.newyorker.com/tech/elements/new-york-citys-bold-flawed-attempt-to-make-algorithms-accountable.

PredPol. "Atlanta Police Chief George Turner Highlights PredPol Usage." PredPol: Blog (May 21, 2014). http://www.predpol.com/atlanta-police-chief-george-turner-highlights-predpol-usage/.

PredPol. "How PredPol Works" [2018]. http://www.predpol.com/how-predictive-policing-works/.

PredPol. "Proven Crime Reduction Results" [2018]. http://www.predpol.com/results/.

Reece, Andrew G., and Christopher M. Danforth. "Instagram Photos Reveal Predictive Markers of Depression." *EPJ Data Science* 6, no. 15 (2017). https://doi.org/10.1140/epjds/s13688-017-0110-z.

Reiman, Jeffrey, and Paul Leighton. *The Rich Get Richer and the Poor Get Prison: Ideology, Class, and Criminal Justice*. New York: Routledge, 2015.

"Results-Driven Contracting: An Overview." Harvard Kennedy School Government PerformanceLab(2016).http://govlab.hks.harvard.edu/files/siblab/files/results-driven _contracting_an_overview_0.pdf.

Reuben, Ernesto, Paola Sapienza, and Luigi Zingales. "How Stereotypes Impair Women's Careers in Science." *Proceedings of the National Academy of Sciences* 111, no. 12 (March 25, 2014): 4403–4408.

"The Rise of Workplace Spying." *The Week*, July 8, 2015. http://theweek.com/ articles/564263/rise-workplace-spying.

Rittel, Horst W. J., and Melvin M. Webber. "Dilemmas in a General Theory of Planning." *Policy Sciences* 4, no. 2 (1973): 155–169.

Robinson, David, and Logan Koepke. "Stuck in a Pattern." *Upturn* (2016). https:// www.teamupturn.org/reports/2016/stuck-in-a-pattern/.

Roller, Emma. "'Victory Can Be a Bit of a Bitch': Corey Robin on the Decline of American Conservatism." *Splinter*, September 1, 2017. https://splinternews.com/victory -can-be-a-bit-of-a-bitch-corey-robin-on-the-dec-1798679236.

Roth, Alvin. "Why New York City's High School Admissions Process Only Works Most of the Time." *Chalkbeat*, July 2, 2015. https://www.chalkbeat.org/posts/ny/2015/ 07/02/why-new-york-citys-high-school-admissions-process-only-works-most-of-the -time/.

Rothstein, Richard. *The Color of Law: A Forgotten History of How Our Government Segregated America*. New York: Liveright, 2017.

Rubin, Joel. "Stopping Crime Before It Starts." *Los Angeles Times*, August 21, 2010. http://articles.latimes.com/2010/aug/21/local/la-me-predictcrime-20100427-1.

Rudder, Christian. "We Experiment On Human Beings!" *The OkCupid Blog* (2014). https://theblog.okcupid.com/we-experiment-on-human-beings-5dd9fe280cd5.

Ryan, John. "After 10-Year Plan, Why Does Seattle Have More Homeless Than Ever?" *KUOW*, March 3, 2015. http://kuow.org/post/after-10-year-plan-why-does-seattle -have-more-homeless-ever.

Sadowski, Jathan, and Frank Pasquale. "The Spectrum of Control: A Social Theory of the Smart City." *First Monday* 20, no. 7 (2015). http://firstmonday.org/article/ view/5903/4660.

Salisbury, Harrison E. *The Shook-Up Generation*. New York: Harper and Row, 1958.

Sampaio, Rafael, and Tiago Peixoto. "Electronic Participatory Budgeting: False Dilemmas and True Complexities." In *Hope for Democracy: 25 Years of Participatory Budgeting Worldwide*, edited by Nelson Dias. São Brás de Alportel, Portugal: In Loco Association, 2014.

San Francisco, City and County of. "Apps: Transportation." *San Francisco Data* [2018]. http://apps.sfgov.org/showcase/apps-categories/transportation/.

Sanders, Lynn M. "Against Deliberation." *Political Theory* 25, no. 3 (1997): 347–376.

Saunders, Jessica, Priscillia Hunt, and John S. Hollywood. "Predictions Put into Practice: A Quasi-experimental Evaluation of Chicago's Predictive Policing Pilot." *Journal of Experimental Criminology* 12, no. 3 (2016): 347–371.

Schiener, Dominik. "Liquid Democracy: True Democracy for the 21st Century" Medium: Organizer Sandbox (November 23, 2015). https://medium.com/organizer -sandbox/liquid-democracy-true-democracy-for-the-21st-century-7c66f5e53b6f.

Schlozman, Kay Lehman, Sidney Verba, and Henry E. Brady. *The Unheavenly Chorus: Unequal Political Voice and the Broken Promise of American Democracy.* Princeton, NJ: Princeton University Press, 2012.

Schmitt, Angie. "How Engineering Standards for Cars Endanger People Crossing the Street." *Streetsblog USA*, March 3, 2017. http://usa.streetsblog.org/2017/03/03/ how-engineering-standards-for-cars-endanger-people-crossing-the-street/.

Schneier, Bruce. *Click Here to Kill Everybody: Security and Survival in a Hyper-connected World.* New York: W. W. Norton, 2018.

Schneier, Bruce. "Data Is a Toxic Asset, So Why Not Throw It Out?" *CNN*, March 1, 2016. http://www.cnn.com/2016/03/01/opinions/data-is-a-toxic-asset-opinion -schneier/index.html.

Schwartz, Paul M., and Daniel J. Solove. "The PII Problem: Privacy and a New Concept of Personally Identifiable Information." *NYU Law Review* 86 (2011): 1814–1894.

Scott, James C. *Seeing Like a State: How Certain Schemes to Improve the Human Condition Have Failed.* New Haven: Yale University Press, 1998.

Seattle, City of. "About the Privacy Program" (2018). http://www.seattle.gov/tech/ initiatives/privacy/about-the-privacy-program.

Seattle, City of. "About the Surveillance Ordinance" (2018). https://www.seattle.gov/ tech/initiatives/privacy/surveillance-technologies/about-surveillance-ordinance.

Seattle, City of. "City of Seattle Privacy Principles" (2015). https://www.seattle.gov/ Documents/Departments/InformationTechnology/City-of-Seattle-Privacy-Principles -FINAL.pdf.

Seattle, City of. "Homelessness Investment Analysis" (2015). https://www.seattle .gov/Documents/Departments/HumanServices/Reports/HomelessInvestment Analysis.pdf.

Seattle/King County Coalition on Homelessness. "2015 Street Count Results" (2015). http://www.homelessinfo.org/what_we_do/one_night_count/2015_results.php.

Semuels, Alana. "The Role of Highways in American Poverty." *The Atlantic*, March 2016. https://www.theatlantic.com/business/archive/2016/03/role-of-highways-in-american-poverty/474282/.

Senseable City Lab. "DriveWAVE by MIT SENSEable City Lab" (2015). http://sense able.mit.edu/wave/.

Shaban, Abdur Rahman Alfa. "Ethiopia Bags a Continental First with $2.2m Smart Parking Facility." *Africanews*, June 15, 2017. http://www.africanews.com/2017/06/15/ethiopia-s-22m-smart-parking-facility-is-africa-s-first/.

Shanker, Ravi, et al. "Autonomous Cars: Self-Driving the New Auto Industry Paradigm." *Morgan Stanley Blue Paper* (November 6, 2013). https://orfe.princeton.edu/~alaink/SmartDrivingCars/PDFs/Nov2013MORGAN-STANLEY-BLUE-PAPER-AUTONOMOUS-CARS%EF%BC%9A-SELF-DRIVING-THE-NEW-AUTO-INDUSTRY-PARADIGM.pdf.

Short, Adrian. "BT InLink in London: Building a Privatised 'Smart City' by Stealth" (December 14, 2017). https://www.adrianshort.org/posts/2017/bt-inlink-london-smart-city/.

Siddle, James. "I Know Where You Were Last Summer: London's Public Bike Data Is Telling Everyone Where You've Been." *The Variable Tree*, April 10, 2014. https://vartree.blogspot.co.uk/2014/04/i-know-where-you-were-last-summer.html.

"Sidewalk Labs." https://www.sidewalklabs.com/.

Sidewalk Labs. "Vision Sections of RFP Submission" (October 17, 2017). https://sidewalktoronto.ca/wp-content/uploads/2017/10/Sidewalk-Labs-Vision-Sections-of-RFP-Submission.pdf.

Singer, Natasha. "Bringing Big Data to the Fight against Benefits Fraud." *New York Times*, February 22, 2015. https://www.nytimes.com/2015/02/22/technology/bringing-big-data-to-the-fight-against-benefits-fraud.html.

Smart, Christopher. "What Do You Like, Don't Like?—Text It to Salt Lake City." *Salt Lake Tribune*, August 20, 2012. http://archive.sltrib.com/story.php?ref=/sltrib/news/54728901-78/lake-salt-text-plan.html.csp.

Smart Columbus. "Smart Columbus Connects Linden Meeting Summary" (2017).

Smith, Jack. "'Minority Report' Is Real—And It's Really Reporting Minorities." *Mic*, November 9, 2015. https://mic.com/articles/127739/minority-reports-predictive-policing-technology-is-really-reporting-minorities.

Solove, Daniel J. *The Digital Person: Technology and Privacy in the Information Age*. New York: New York University Press, 2004.

Sorokanich, Bob. "New York City Is Using Data Mining to Fight Fires." *Gizmodo* (2014). https://gizmodo.com/new-york-city-is-fighting-fires-with-data-mining-1509 004543.

Speck, Jeff. "Autonomous Vehicles & the Good City." Lecture at United States Conference of Mayors, Washington, DC, January 19, 2017. https://www.youtube.com/watch?v=5AELH-sI9CM.

Spurr, Ben. "Toronto Plans to Test Driverless Vehicles for Trips to and from Transit Stations." *The Star*, July 3, 2018. https://www.thestar.com/news/gta/2018/07/03/toronto-plans-to-test-driverless-vehicles-for-trips-to-and-from-transit-stations.html.

Sullivan, Christopher M., and Zachary P. O'Keeffe. "Evidence That Curtailing Proactive Policing Can Reduce Major Crime." *Nature Human Behaviour* 1 (2017): 730–737.

Sweeney, Latanya. "Simple Demographics Often Identify People Uniquely." Carnegie Mellon University, Data Privacy Working Paper 3, 2000.

Tachet, Remi, Paolo Santi, Stanislav Sobolevsky, Luis Ignacio Reyes-Castro, Emilio Frazzoli, Dirk Helbing, and Carlo Ratti. "Revisiting Street Intersections Using Slot-Based Systems." *PloS One* 11, no. 3 (2016). https://doi.org/10.1371/journal.pone .0149607.

Tadayon, Ali. "Oakland to Require Public Approval of Surveillance Tech." *East Bay Times*, May 2, 2018. https://www.eastbaytimes.com/2018/05/02/oakland-to-require -public-approval-of-surveillance-tech/.

Tauberer, Joshua. "So You Want to Reform Democracy." Medium: Civic Tech Thoughts from JoshData (November 22, 2015). https://medium.com/civic-tech-thoughts -from-joshdata/so-you-want-to-reform-democracy-7f3b1ef10597.

TechCrunch. "Taking a Ride in Delphi's Latest Autonomous Drive" (2017). https:// www.youtube.com/watch?v=wWdVfGlBqzE.

Tett, Gillian. "Mapping Crime–or Stirring Hate?" *Financial Times*, August 22, 2014. https://www.ft.com/content/200bebee-28b9-11e4-8bda-00144feabdc0.

"Textizen." https://www.textizen.com/.

Ticoll, David. "Driving Changes: Automated Vehicles in Toronto." Discussion paper, Munk School of Global Affairs, University of Toronto (2015). https://munkschool .utoronto.ca/ipl/files/2016/03/Driving-Changes-Ticoll-2015.pdf.

Tockar, Anthony. "Riding with the Stars: Passenger Privacy in the NYC Taxicab Dataset." *Neustar Research*, September 15, 2014. https://research.neustar.biz/2014/09/15/ riding-with-the-stars-passenger-privacy-in-the-nyc-taxicab-dataset/.

Tufekci, Zeynep. *Twitter and Tear Gas: The Power and Fragility of Networked Protest.* New Haven: Yale University Press, 2017.

United States Conference of Mayors. "Cities of the 21st Century: 2016 Smart Cities Survey" (January 2017). https://www.usmayors.org/wp-content/uploads/2017/02/2016SmartCitiesSurvey.pdf.

U.S. Department of Transportation. "Smart City Challenge: Lessons for Building Cities of the Future" (2017). https://www.transportation.gov/sites/dot.gov/files/docs/Smart City Challenge Lessons Learned.pdf.

Vaithianathan, Rhema. "Big Data Should Shrink Bureaucracy Big Time." *Stuff*, October 18, 2016. https://www.stuff.co.nz/national/politics/opinion/85416929/rhema-vaithianathan-big-data-should-shrink-bureaucracy-big-time.

Vargas, Claudia. "City Settles Gun Permit Posting Suit." *Philadelphia Inquirer*, July 23, 2014. http://www.philly.com/philly/news/local/20140723_City_settles_gun_permit_suit_for__1_4_million.html.

Vitale, Alex S. *The End of Policing*. London: Verso, 2017.

Wakabayashi, Daisuke. "Uber's Self-Driving Cars Were Struggling Before Arizona Crash." *New York Times*, March 23, 2018. https://www.nytimes.com/2018/03/23/technology/uber-self-driving-cars-arizona.html.

Wang, Tong, Cynthia Rudin, Daniel Wagner, and Rich Sevieri. "Finding Patterns with a Rotten Core: Data Mining for Crime Series with Cores." *Big Data* 3, no. 1 (2015): 3–21. http://doi.org/10.1089/big.2014.0021.

The War Room. Hosted by Jennifer Granholm, Current TV, January 16, 2013. Posted as "PredPol on Current TV with Santa Cruz Crime Analyst Zach Friend" (2013). https://www.youtube.com/watch?v=8uKorOnfsdQ.

Washington, Ken. "A Look into Ford's Self-Driving Future." Medium: Self-Driven (February 3, 2017). https://medium.com/self-driven/a-look-into-fords-self-driving-future-5aae38ee2059.

Weisel, Deborah Lamm. "Burglary of Single-Family Houses." *U.S. Department of Justice, Office of Community Oriented Policing Services, Problem-Oriented Guides for Police Series No. 18*, 2002. http://www.popcenter.org/problems/pdfs/burglary_of_single-family_houses.pdf.

Wexler, Rebecca. "Life, Liberty, and Trade Secrets: Intellectual Property in the Criminal Justice System." *Stanford Law Review* 70 (2018): 1343–1429.

What Works Cities. "Tackling Homelessness in Seattle" (2017). https://www.youtube.com/watch?v=dzkblumT4XU.

White, Ariel, and Kris-Stella Trump. "The Promises and Pitfalls of 311 Data." *Urban Affairs Review* 54, no. 4 (2016): 794–823. https://doi.org/10.1177/1078087416673202.

White House Office of the Press Secretary. "FACT SHEET: Administration Announces New 'Smart Cities' Initiative to Help Communities Tackle Local Challenges and Improve City Services" (September 14, 2015). https://obamawhitehouse.archives .gov/the-press-office/2015/09/14/fact-sheet-administration-announces-new -smart-cities-initiative-help.

White House Office of the Press Secretary. "FACT SHEET: Launching the Data-Driven Justice Initiative: Disrupting the Cycle of Incarceration" (June 30, 2016). https:// obamawhitehouse.archives.gov/the-press-office/2016/06/30/fact-sheet-launching -data-driven-justice-initiative-disrupting-cycle.

Winner, Langdon. *The Whale and the Reactor: A Search for Limits in an Age of High Technology.* Chicago: University of Chicago Press, 1986.

Winston, Ali. "Palantir Has Secretly Been Using New Orleans to Test Its Predictive Policing Technology." *The Verge,* February 27, 2018. https://www.theverge.com/2018/ 2/27/17054740/palantir-predictive-policing-tool-new-orleans-nopd.

Wylie, Bianca. "Debrief on Sidewalk Toronto Public Meeting #2—Time to Start Over, Extend the Process." Medium (May 6, 2018). https://medium.com/@biancawylie/ sidewalk-toronto-public-meeting-2-time-to-start-over-extend-the-process-a0575b 3adfc3.

Zimmer, John. "The Third Transportation Revolution." Medium: The Road Ahead (September 18, 2016). https://medium.com/@johnzimmer/the-third-transportation -revolution-27860f05fa91.

Zimmer, John, and Logan Green. "The End of Traffic: Increasing American Prosperity and Quality of Life." Medium: The Road Ahead (January 17, 2017). https:// medium.com/@johnzimmer/the-end-of-traffic-6d255c03207d.

Zuckerberg, Mark. "Facebook's Letter from Mark Zuckerberg—Full Text." *The Guardian,* February 1, 2012. https://www.theguardian.com/technology/2012/feb/01/facebook -letter-mark-zuckerberg-text.

Index